Production Technology for Fruit and Plantation Crops

The Authors

Dr Mukesh Kumar completed his Doctorate in Horticulture from the College of Agriculture, CCS HAU, Hisar in 2006 and joined as Assistant Professor and equivalent in 2012. During his scientific career, he worked in teaching (UG and PG), research and extension at different stations of CCS HAU, Hisar. He guided 10 PG and PhD students. He is a prolific writer and has written one book, about 50 research papers and more than 120 abstracts, book chapters, popular articles *etc.*, in national and international journals and magazines. He has two gold medals, three recommended technologies and one patent that has also been granted to him by the Govt of India.

Dr Jeet Ram Sharma completed his post-graduation from the College of Agriculture, CCS HAU, Hisar in 1987 and joined university services in 1990. During his scientific career, he worked in teaching (UG and PG), research, extension at different stations of CCS HAU, Hisar and has completed his Ph.D. (in-service). He has served 13 years in various administrative posts. He guided 17 PG and PhD students. He is a creative writer and has written one book, about 105 research papers and more than 170 abstracts, book chapters, popular articles *etc.*, in national and international journals and magazines. He recommended three technologies and two hybrids are in the pipeline for recommendation.

Production Technology for Fruit and Plantation Crops

— Authors —

Dr Mukesh Kumar

Assistant Professor (Horticulture)
CCS Haryana Agricultural University
College of Agriculture, Bawal (Rewari)

Dr Jeet Ram Sharma

Director of Research
(Professor Horticulture)
CCS Haryana Agricultural University, Hisar

2025

Daya Publishing House®
A Division of
Astral International Pvt. Ltd.
New Delhi – 110 002

ISBN: 978-93-5919-730-2
UPN No.: CCSHAU/PUB#23-104

Published by : **Daya Publishing House®**
A Division of
Astral International Pvt. Ltd.
– ISO 9001:2015 Certified Company –
4736/23, Ansari Road, Darya Ganj
New Delhi-110 002
Ph. 011-43549197, 23278134
E-mail: info@astralint.com
Website: www.astralint.com

Printed at : **Replika Press Pvt Ltd,**

Prof. B.R. Kamboj

Vice Chancellor
CCS Haryana Agricultural University
Hisar – 125004, India

Foreword

Fruit and plantation crops are important sectors for diversification in agriculture. These are helpful to increase production, generate employment and enhance exports. India is the second largest producer of fruits in the world after China. Globally, India is at the first position in the production of banana, mango, pomegranate, lime, lemon, papaya, coconut and cashew, and productivity of grape, banana, cassava and papaya. Fruit and plantation crops are more resilient to changes in weather conditions. This sector enables the large population to enjoy a healthy and balanced diet. Although these are required in very minute quantities, they are absolutely essential for the maintenance of health. Fruit and vegetables supply several vitamins. The deficiency of any vitamin for a considerable period may lead to a diseased state or disorder conditions.

Per capita availability of fruits for good health should be 230 g/day. Many developed and underdeveloped countries are unable to meet this requirement. In India, per capita availability of fruit is about 207.9 g per day. So, there is great scope for the utilization and consumption of fruits within the country to meet our export demand. Production and productivity can be enhanced by the adoption of quality planting material, efficient utilization of land and water, knowledge about production technology, infrastructure and assured market rate.

This book is enriched by including technical information related to the production of fruit and plantation crops, such as importance and scope of the fruit and plantation crop industry in India; high density planting and use of rootstocks, production technologies for cultivation of major fruits – mango, ber, aonla, date palm, sapota, plum, citrus, grape, guava, litchi, papaya, pear, peach and minor fruits – pineapple, pomegranate, jackfruit, strawberry, nut crops; plantation crop – coconut, areca nut, cashew, tea, coffee and rubber *etc.*

I appreciate the efforts made by the authors Dr Mukesh Kumar and Dr Jeet Ram Sharma to compile this informative book for the course Production technology for fruit and plantation crops as per ICAR norms. I hope this book will be helpful to the students of UG and PG classes for preparation for different examinations, researchers, teachers and extension workers to empower their knowledge and contribute to the nation's development and doubling farmers' income.

Prof. B.R. Kamboj

Dr. S.K. Pahuja
Dean, College of Agriculture
CCS HAU, Hisar -125004
Haryana, India

Foreword

Fruits are becoming one of the essential components of the Indian diet. These also help in nutritional security, employment generation and economic development, environment protection and now in international trade and overall growth of the country. Fruit crops are grown in an area of 7.06 m ha, which contributes 107.5 m MT production. However, food grain crops cover 130.2 m ha area and contribute 315.6 m MT production. It is indicated from the figure that about 5.4 per cent of the area under fruit crops contributes 34 per cent of production. The production of horticultural produce has increased 7-fold since 1950, which ensures nutritional security and employment opportunities. Export growth of fresh fruit and vegetables in terms of value is 14 per cent and of processed fruit and vegetables is 16.27 per cent.

Apart from yield, income and employment, this sector also provides nutritional benefits. At the same time, fruit crops are good sources of vitamins and minerals, so called as protective food and are an integral part of the human diet. Some fruits like aonla, bael, pomegranate, jamun, karonda, guava, kokum, ber *etc.*, having medicinal value (aonla, triphala and *chawanprash*), are recommended for disorders like night blindness (papaya), diarrhoea (citrus juice), vitamin C and digestion.

The book is enriched by including technical information related to the production of fruit and plantation crops, such as the importance and scope of the fruit and plantation crop industry in India; high density planting and use of rootstocks, production technologies for the cultivation of major fruits – mango, ber, aonla, date palm, sapota, plum, citrus, grape, guava, litchi, papaya, pear, peach and minor fruits – pineapple, pomegranate, jackfruit, strawberry, nut crops; plantation crop – coconut, areca nut, cashew, tea, coffee and rubber *etc.*

I appreciate the efforts made by the authors Dr Mukesh Kumar and Dr Jeet Ram Sharma to compile the information for the course Production technology for fruit and plantation crops as per ICAR norms. I believe that this book will be helpful for the UG and PG students in Agriculture, Horticulture, Forestry and for competitive examinations like JRF/SRF/NET of agriculture/horticulture and its allied branches. It will also be helpful for researchers, teachers and extension workers to empower their knowledge.

Dr. S.K. Pahuja

Dr. Naresh Kaushik

Principal
CCS HAU College of Agriculture,
Bawal-123501, Rewari, India

Foreword

India is the second largest producer of fruit and vegetables at the global level. Production of fruits per unit area is higher than other crops. The average yields of fruit crops like papaya, banana and grapes is 10 to 15 times higher than that of crops. Due to higher returns per unit area from the horticultural crop, the person engaged in this activity has a higher social status. More production, productivity and market value improve the farmer's social status and decrease unemployment by engaging more labour in horticultural crops as compared to field crops. Fruit crops provide employment for small farmers and agricultural labour throughout the year. One hectare of fruit production generates 860 man-days per annum as against 143 man-days for cereal crops and the crops like grapes, banana and pineapple need 1000- 2500 man days per annum.

Barren land can be efficiently used for growing arid and semi-arid fruit crops *e.g.* ber, aonla, phalsa, jamun, cashew, custard apple *etc.,* which can be grown successfully on poor, shallow, undulated soils considered unsuitable for growing grain/agronomic crops.

Fruit and plantation crops are directly used by people for food, medicinal purposes or for aesthetic gratification. Fruits are used to worship Gods, mango leaves tied at the entrance of the house are considered auspicious. Fruit crops provide a better microclimate to the surrounding area. The perennial nature of fruits provides shelter to birds, small animals, reptiles and other microbes and add diversity to the ecology. These also help to increase greenery, carbon sequestration, maintain biodiversity and protect the environment.

The book is enriched by including technical information related to production of fruit and plantation crops, such as the importance and scope of fruit and plantation crop industry in India; high-density planting and use of rootstocks, production technologies for cultivation of major fruits – mango, ber, aonla, date palm, sapota, plum, citrus, grape, guava, litchi, papaya, pear, peach and minor fruits – pineapple, pomegranate, jackfruit, strawberry, nut crops; plantation crop – coconut, areca nut, cashew, tea, coffee and rubber *etc.*

I appreciate the efforts made by the authors Dr Mukesh Kumar and Dr Jeet Ram Sharma to compile the information for the course Production technology for fruit and plantation crops as per ICAR norms. I believe that this book will be helpful for the UG and PG students in Agriculture, Horticulture, Forestry and for competitive examinations like JRF/SRF/NET of agriculture/horticulture and its allied branches. It will also be helpful for researchers, teachers and extension workers to empower their knowledge.

Dr. Naresh Kaushik

Preface

The book on Production technology for fruit and plantation crops is prepared as per the course curriculum recommended by CCS HAU, Hisar, SAUs and ICAR. The book is enriched by including technical information related to the production of fruit and plantation crops, such as the importance and scope of fruit and plantation crop industry in India; high density planting and use of rootstocks, production technologies for cultivation of major fruits – mango, ber, aonla, date palm, sapota, plum, citrus, grape, guava, litchi, papaya, pear, peach and minor fruits – pineapple, pomegranate, jackfruit, strawberry, nut crops; plantation crop – coconut, areca nut, cashew, tea, coffee and rubber *etc.* There are more than 100 fruit crops, but the authors have covered as per the recommendation of ICAR for UG students. The topics are covered with pictorial representation for understanding of the concept.

We hope this book will serve as a handy guide for students of graduation in horticulture/agriculture, post-graduation and preparation for JRF/SRF/NET and other competitive examinations. It will also be helpful for researchers, extension workers, orchardists and others interested in fruit culture.

To write this book, we have consulted several books, research papers, reports and web content from different institutions in order to bring in the most recent and relevant information. The chemicals/pesticides banned or may be banned in future by the Government of India should not be recommended or used. Efforts have been made to cover all the concepts on different topics but there are always chances of improvement in each and every document. Despite, our best efforts, we do not claim this book is complete in all sense. It may have some shortcomings, the authors will be really grateful to receive suggestions and criticism from the readers for improvement and incorporation in forthcoming issues.

We are thankful to the University authority and others who helped directly or indirectly in the preparation of this book. We also pay tribute to our lovely family for their inspiration and cooperation during the preparation of this manuscript.

Mukesh Kumar

J. R. Sharma

Contents

Chapter 1

Importance and Scope of Fruit and Plantation Crop Industry in India

History of Fruit Cultivation is as old as Civilization

Fruit have their references in *Vedas, Puranas, Upnishads etc.*

- ☆ The oldest fruit plant is Date palms (reference dates back to 7000 B.C.).
- ☆ Pomegranate (3500 B.C.), Grapes (2440 B.C.).
- ☆ Mango, Banana and Coconut (2000 B.C.).
- ☆ Peach and Almond (1300 B.C.), Olive (100 B.C.).
- ☆ The first book exclusively on litchi cultivation was published in China in 1056 A.D.

Horticulture: The term is derived from the Latin word '*hortus*', which means garden and '*cultura*', which means cultivation. It deals with the cultivation of garden crops, which include fruit, vegetables and ornamental crops.

Horticulture: It is a branch of agricultural sciences that deals with intensively cultured plants directly used by people for food, medicinal purpose or aesthetic gratification.

Modern Horticulture: It is a branch of agricultural sciences which deals with the production, utilization, improvement and post-harvest management of fruits, vegetables, ornamental plants, spices, plantations, medicinal and aromatic crops.

Branches of Horticulture

- Fruit Science (Pomology)
- Vegetable Science (Olericulture)
- Floriculture and Landscape Architecture
- Plantation, Spices, Medicinal and Aromatic Crops

Fruit Science (Pomology)

Pomology is a branch of horticulture which deals with fruits. '*Poma*' is derived from Greek, which means fruits, and was later subsequently transferred to '*Pome*' a Latin word that means fruits and '*logos*' means study.

Branches of Pomology

a) **Basic Pomology**: Deals with the fundamental aspects of fruit production like methods of plantation, irrigation, training and pruning, water, nutrients, insect pest and disease management, use of plant growth regulators *etc.*

b) **Commercial Pomology**: It deals with commercial fruit production technology.

c) **Systematic Pomology**: It deals with nomenclature and classification of fruit plants into kingdom, order, suborder, class, family, genus and species.

Vegetable Science (Olericulture)

The '*oleris*' is derived from the Latin word which means pot herb, and '*cultura*' means cultivation. Olericulture is a branch of horticulture which deals with vegetable production. Basically, olericulture is the art and science of the so-called truck crops.

Floriculture and Landscaping Architecture

Floriculture is the science of flower production. It deals with all the operations and practices used to take care of all such plants, which are cultivated exclusively for their flowers and foliage.

Landscape architecture is the appearance of land as it appears to the eyes. Landscape gardening is the art of beautifying a piece of land using garden design, methods and plant materials.

Plantation, Spices, Medicinal and Aromatic Crops

Plantation crops: Perennial crops grown at large scale and their products are used after processing, *e.g.* coconut, areca nut, cashew nut, tea, coffee, rubber *etc.*

Species: Different parts of plants are used to add flavour to the food, *e.g.* black pepper, cardamom, clove, chilli, cumin *etc.*

Medicinal crops: Medicinal crops have different compounds in plants or plant parts are used in pharmaceutical industries *e.g.* aonla, bael, isabgol *etc.*

Aromatic Plants: Volatile oil or compound found in various plant or plant parts, *e.g.* rose, mint, lemongrass *etc.*

Fact About Fruit Production

- ☆ India is the 7th largest country in the world with a geographical area 328.73 m ha, 1st in the population (140 crores) and 2nd in fruit production after China.
- ☆ The total arable land in India is 144 m ha.
- ☆ In India, about 55-60 per cent of the population directly or indirectly depends on agriculture and its allied activities.
- ☆ India ranks 1st in productivity of grapes, *i.e.* 20.81 MT/ha and 3rd in productivity of Banana, *i.e.* 35.87 MT/ha.
- ☆ The area under horticultural crops in India is 28.04 m ha and production 347.18 m MT
- ☆ The area under fruit crops in India is 7.06 m ha, production is 107.51 m MT and productivity is 15.22 MT.
- ☆ The area under plantation crops in India is 4.29 m ha, production is 15.76 m MT and productivity is 3.67 MT.
- ☆ Production share of banana is (32.1 %), mango (19.3 %), citrus (13.7 %), papaya (4.96 %) and guava (5.14 %) out of total fruit production.
- ☆ The area under fruit crops in Haryana is 69140 ha, with a production of 1010490 MT and productivity of 14.61 MT.

Table 1.1: Position of India in Fruit Production at the World Level

Fruit crops	*Rank at world level*	*Share in world production*
Mango/guava/mangosteen	1st	40.4 per cent
Papaya	1st	44.4 per cent
Banana	1st	25.7 per cent
Lemon, lime	1st	17.2 per cent

Table 1.2: Leading Countries in Area and Production of Fruits

Countries	*Area (m ha)*	*Countries*	*Production (m MT)*
China	16.59	China	272.08
India	6.98	India	90.89
Brazil	2.33	Brazil	39.69
Philippines	1.68	USA	27.11
Spain	1.63	Mexico	21.43

Table 1.3: Leading States in Area and Production of Fruits

State	*Area share (%)*	*State*	*Production share (%)*
Maharashtra	11.41	Andhra Pradesh	18.17
Andhra Pradesh	10.76	Uttar Pradesh	10.95
Uttar Pradesh	7.30	Maharashtra	10.09
Gujarat	6.51	Gujarat	9.31
Madhya Pradesh	5.63	Madhya Pradesh	7.83
Karnataka	5.61	Karnataka	6.62

Importance and Scope of Fruit Production

1. Economic Importance

A well-maintained orchard brings better returns per unit area than field crops. Although initial expenses involved in the establishment of an orchard are much more than raising field crops, this can be compensated to a large extent by taking inter-crop during the initial years of planting. High-value crops with a high potential for value addition make a higher contribution to GDP, *i.e.* 30 per cent from 14 per cent of the total cropped area under horticultural crops. The output share of horticultural crops is 40 per cent of total agricultural output.

a) **High productivity:** Production per unit area from fruit crops is higher than other crops. The average yield of fruit crops like papaya, banana and grapes is 10 to 15 times more than that of field crops.

b) **High net profit:** No doubt the establishment cost of a new orchard is higher than field crops, but it is compensated by increased production, its value, productivity and net profit, *e.g.* the yield of wheat/ground nut/*ragi* is about 3.0 to 4.0 MT/ha valued Rs. 25000/- to 35,000/- as compared to the production of grapes/mango/banana which is about 20-40 MT/ha valued Rs 1.5-2.5 lakh, which is about 5 to 6 times more than field crops.

c) **Provide input/raw material for agro-based industries:** Fruit is one of the essential inputs for the different fruit processing industries and other agro-industries like canning and preservation of fresh fruits like aonla, bael, coir industries (coconut husk), pharmaceutical industry (aonla, bael, papaya, jamun), transportation and packaging industries (banana leaf and wood of other fruits) *etc.* Many horticultural products and their byproducts are used as raw materials for many industries, *e.g.* rubber, coir (coconut) and sago (tapioca) industries.

d) **Efficient utilization of resources:** All the natural and artificial farm resources may be utilized throughout the year because fruit crops are perennial in nature so farm labour, machinery, land and water are required throughout the year as compared to field crops.

e) **Utilization of waste and barren lands for production:** Most of the fruit crops are perennial in nature and require good quality water and soil for production. However, some arid and semi-arid fruit crops are of hardy nature *e.g.* ber, aonla, phalsa, jamun, cashew, custard apple *etc.*, which can be grown successfully on poor, shallow, undulated soils which are considered unsuitable for growing grain/agronomic crops.

f) **Foreign exchange:** Export of fresh fruits, vegetables and their processed products are in great demand in foreign countries and thus are a source of earning foreign exchange, *e.g.* fresh fruits and vegetables – apple, mango, grapes, mushroom, onion, potatoes *etc.*, flowers-cut flowers (roses), potted and foliage plants, plantation-cashewnut, tea, coffee, coconut, arecanut, *etc.*

2. Important for Nutritional Security/Protective Food

Horticultural crops are rich sources of vitamins and minerals, so they are called protective food. Considering the value of these crops to human health, 230 g of fruits and 400 g of vegetables per capita per day are recommended for good human health.

$$\textbf{\textit{Per capita availability of fruits (g)}} = \frac{Total\ Production\ (g) - (Post\ harvest\ Losses + Export)}{Total\ Population\ x\ 365}$$

a) Carbohydrates, fats and protein

Carbohydrates and fats provide energy to the human body, while protein and fat are essential as these play an important role in the growth and development of the human being.

Table 1.4: Composition of Carbohydrates, Fats and Protein in Fruits

Fruit crops	*Carbohydrates (%)*	*Fat (%)*	*Protein (%)*
Almond	10.5	58.7 (2nd)	20.8 (2nd)
Apricot (dry)	72.8 (2nd) *	0.7	1.7
Avocado	0.8	22.8	1.7
Banana	36.4	0.2	1.3
Bael	30.6	0.2	1.3
Ber	13.6	0.1	0.9
Cashewnut	22.3	46.9 (3rd)	21.2 (1st)
Dates	67.3 (3rd)	0.2	3.0
Karonda	67.1 (4th)	9.6	2.3
Raisins (dried grapes)	77.3 (1st)	0.2	2.0
Walnut	11.0	64.5 (1st)	15.6 (3rd)

* Ranks.

b) Calorific value

According to the National Research Council, Food and Nutrition Board, men and women aged between 23 to 50 years require about 2800 and 2200 calories a day, respectively, to maintain weight. The nutrient requirement varies with gender, age, height, weight, physical condition, activity level and climatic conditions where they live, *e.g.* dates supply 281 calories/100 g, avocado - 190 calories/100 g, banana - 95 calories/100 g, peach - 94 calories/100 g.

c) Vitamins

Vitamins are important constituents of fruits and vegetables and are an essential part of the human diet. These are required in very small amounts but are essential for the maintenance of health. The deficiency of any vitamin may lead to disease or disorder. The following vitamins are supplied by fruit and vegetables.

i) **Vitamin-A (Carotene):** It is essential for average growth, reproduction and maintenance of health and vigour. It protects against colds, influenza and prevents night blindness. Its deficiency results in cessation of development in young children, night blindness, drying up of tear glands in the eyes, eruption of the skin (rashes on the skin) and brittleness of the teeth.

   Sources: Mango (4800 IU/100 g), papaya (2020 IU/100 g), dates, jackfruit, walnut *etc* (one IU biologically equivalent to 0.3 mcg retinol, mcg-micrograms).

ii) **Vitamin B_1 (Thiamine):** Tones the nervous system and helps in proper functioning of the digestive tract. Its deficiency causes beri-beri, paralysis, loss of sensitivity of skin, enlargement of the heart, loss of appetite, loss of weight and fall in body temperature.

   Sources: Cashew nut (630 mg/100 g), walnut (450 mg/100 g), dry apricot (217 mg/100 g), orange, pineapple, jackfruit, almond, banana *etc.*

iii) **Vitamin B_2 (Riboflavin):** It is required for body growth and health of the skin. The deficiency of this vitamin causes sore throat, anorexia cataract, loss of appetite, body weight and the development of a swollen nose.

   Sources: Bael (1191 mg/100 g), papaya (250 mg/100 g), litchi (122 mg/100 g), banana, apricot, pomegranate, pear *etc.*

iv) **Vitamin C (Ascorbic Acid):** It promotes general health and healthy gums, prevents scurvy disease, which is characterized by pain in the joints and swelling of limbs (rheumatism), bleeding of gums, tooth decay and keeps the blood vessels in good condition.

   Sources: Barbados cherry (1500-4000 mg/100 g), aonla (600-1200 mg/100 g), guava (220-300 mg/100 g), ber (120 mg/100 g), citrus (50-60 mg/100 g), strawberry, pineapple *etc.*

d) Minerals

A large number of minerals are present in the human body. Bones and teeth are mainly made up of calcium, magnesium and phosphorus. Iron is an essential constituent of blood, whereas iodine regulates the functioning of hormones (thyroxine). Minerals like zinc, molybdenum and manganese are activators of certain enzymes. In addition to these, many of these elements are also required for adequate growth of tissues.

Table 1.5: Minerals Composition in Fruits

Fruit crops	*Calcium (%)*	*Phosphorus (%)*	*Iron (mg/100 g)*
Almond	0.23 (1st)	0.39	3.47
Aonla	0.05	0.02	1.20
Apple	0.01	0.01	1.68
Ber	0.03	0.04	1.14
Cashew nut	0.05	0.45 (1st)	5.00 (3rd)
Date palm	0.07	0.08	10.60 (2nd)
Grapes	0.005	0.02	0.40
Guava	0.01	0.04	1.04
Karonda	0.16 (3rd)	0.06	39.70 (1st)
Mango	0.007	0.02	0.25
Papaya	0.01	0.01	0.27
Peach	0.006	0.03	1.70
Raisins	0.10	0.08	4.00
Walnut	0.10	0.38 (2nd)	4.80
Litchi	0.21 (2nd)	0.30 (3rd)	1.00

e) Enzymes

Enzymes are required for different metabolic activities in the body.

Sources: Papaya for papain and Pineapple for bromelain.

f) Fibre and roughages (Cellulose and pectin)

Roughages are required for digestion and prevention of constipation.

Sources: Fruit are low in fibre content. However, guava (6.90 %), anola (5.20 %) and pomegranate (5.10 %) are better sources than other fruits. Leafy vegetables are rich in fibre content.

g) Medicinal value

Some fruits like aonla, bael, pomegranate, jamun, karonda, guava, kokum, ber *etc.*, have medicinal value and these are recommended as remedies for the following disorders:

- Papaya is a rich source of vitamin A, which helps to reduce night blindness.
- Citrus juice reduces acute diarrhoea.
- Aonla is rich in vitamin C and an important constituent in *triphala* (*chawanprash*) and helps in digestion.

3. Social and Religious Importance

Horticulture is a part of civilization also. Fruits are being used for the worship of Gods. Tying of mango leaves at the entrance of the house is considered auspicious. Due to higher returns per unit area from the horticultural crop, the person engaged in this activity has a higher social status.

4. Employment Opportunity

Horticultural crops provide employment for small farmers and agricultural labour throughout the year. One hectare of fruit production generates 860 man-days per annum as against 143 man-days for cereal crops and the crops like grapes, bananas and pineapple need 1000- 2500 man-days per annum.

5. Other Importance

- Horticultural trees work in a similar way to forest trees in maintaining the ecosystem.
- Fruit crops provide a better microclimate.
- These are perennial in nature, so they provide shelter to birds, small animals, reptiles and other microbes as well as add diversity to the ecology.
- These provide impetus and a calm environment for thinkers, writers, poets and analysts to convert their thinking into words.
- Add values to the all living entity.

We can say the horticultural crops are better for health, wealth, happiness and hygiene.

Scope of Horticulture

In brief, it can be stated that horticulture has excellent scope for the following reasons:

- Horticultural crops are grown in different climatic zones, so variability can be exploited in different agro-climatic conditions.
- It is helpful to meet the nutritional demand and other needs of the increasing population for fruits, vegetables, flowers, spices, beverages *etc.*
- Meet the requirement of raw materials for different processing industries.
- Substitute imports and earn foreign exchange by increasing exports.

- More production, productivity and market value, which improves the social status of the farmers and decrease the unemployment problem by engaging more labour in horticultural crops as compared to field crops.
- It helps to increase greenery, carbon sequestration, maintain biodiversity and protect the environment.

Constraints

A) Production Constraints

- Inadequate supply of quality planting material.
- Low density in existing plantation of orchard.
- Lack of technical knowledge and skills among growers.
- Untrained labour for different operations.
- Poor tree canopy management/tree architecture.
- Small /decreasing land holding.
- Fruit crops have a long gestation period.
- Pre-dominated old and/or senile unproductive orchards.
- Poor quality produce due to low-grade varieties and mismanagement.
- Non-adoption of appropriate agro-techniques and good management practices.

B) Protection Constraints

- Indiscriminate use of pesticides and fungicides for immediate control of major pests and diseases and lack of integrated pest management (IPM) practice.
- Few unsolved national problems are hindrance in fruit production *e.g.* mango malformation, guava wilt, citrus decline, citrus granulation, biennial bearing *etc.*

C) Post-harvest/ Handling Constraints

- Early harvesting of fruit before attaining maturity.
- Lack of basic post-harvest handling infrastructure facilities.
- Non-adoption of post-harvest technologies.
- Absence of pre-cooling, cold chain and cold storage.
- Non-availability of processing-specific varieties.

D) Policy Support

- Lack of suitable laws in the Nursery Act and contract farming.
- Lack of credit support to farmers.
- Lack of trainings for farmers/growers.

- ✰ Formulation of standards for planting materials, produce quality, phytosanitary, packaging standards, as well as standards for pesticide residues.
- ✰ Preparation of standards on quality system *viz.*, hazard analysis critical control points (HACCP), good manufacturing practices (GMP), good hygiene and sanitation practices.
- ✰ Lack of market facilities, market intelligence and market information services (MIS).

Strategies

1. Increase availability of good quality seeds and planting material.
2. Rejuvenate the old orchard.
3. Optimum utilization of land space by adopting high-density planting.
4. Adopt soil health management/nutrient management practices.
5. Leaf nutrient analysis techniques are to be followed for diagnostic and recommendation integrated system (DRIS).
6. Efficient management of inputs like nutrients, irrigation water (using plastic mulching and drip irrigation). Use greenhouses, low tunnels, shading nuts, film for drying produce, propagation in nurseries and water conservation through pond lining.
7. Adopt biocontrol measures for major insect-pests and diseases to reduce pesticide toxicity. The promotion of organic farming should be given due attention.
8. Use of growth-promoting chemicals for increasing productivity *e.g.* use of gibberellins in grapes and paclobutrazol (Cultar) in mango.
9. Adopt precision farming in horticulture. Precision farming is the efficient management of resources through location-specific high-technology interventions.
10. Follow proper packaging and storage standards to avoid post-harvest losses.
11. Development/utilization of cold chain facilities.
12. Establishment of processing industries in intensive fruit growing areas.
13. Make fruit production-oriented credit policies and timely release of assistance as per the seasonal requirement.
14. Formulation of standards on quality planting material, phytosanitary, as well as standards for pesticide residue in fruit crops.
15. Training to growers and use a single window system to solve growers' problems.

Hi-tech Horticulture

Such technologies can be defined as those which are modern, less environment -dependent, capital intensive and have the capacity to improve productivity and quality. These includes the use of GM (genetically modified) crops/varieties, HDP (high-density planting), micropropagation, integrated nutrient, water and pest management, use of plastics mulching, precision farming, protected cultivation, organic farming, use of modern immunodiagnostic techniques for quick detection of viral diseases and hi-tech post-harvest technologies, including cold chain.

Plantation Crops

These are the crops that are grown on a large scale and their produce cannot be consumed/used without processing. These crops provide employment to labour throughout the year. Plantation crops are grown on a large scale and are also called as estate crops. *e.g.* tea, coffee, cashewnut, coconut, rubber, arecanut *etc.*

Major characteristics of plantation crops are:

i. These are perennial in nature.
ii. Grown on a large scale.
iii. Consumed after processing.
iv. Mostly grown under tropical and subtropical conditions.
v. Provide employment throughout the year.
vi. The system of cultivation is uniform.

Table 1.6: Difference Between Plantation and Fruit Crops

Sr.No.	*Plantation crops*	*Fruit crops*
1.	These are mostly grown under tropical conditions.	These are grown under tropical, subtropical and temperate conditions.
2.	The requirement of training and pruning is very rare.	Most of the crops require regular training and pruning.
3.	These are consumed after processing.	These are consumed fresh as well as in processed form.
4.	Fruit/produce are less perishable in nature.	Their produce or fruit are more perishable in nature.
5.	The export potential of plantation crops is more.	The export potential is in lower quantity. Mostly preserved fruits are exported.
6.	These are used for beverages, oils and medicinal purposes.	These are mostly used as protective food because they supply vitamins, minerals, fibers and antioxidants.

On the basis of use plantation crops are classified as under

- Industrial use: Rubber.
- Nut crops: Coconut, arecanut and cashewnut.

- Beverages: Cocoa, tea and coffee.
- Masticatory: Betel vine and arecanut.

Production Statistics

- The total area under plantation crops in India is 4.29 m ha and production is 15.76 m MT with an average productivity of 3.67 MT/ha.
- The area shared under plantation crops is about 16.86 per cent of horticultural crops.
- The production share of plantation crops is about 5 per cent of horticultural crops.

Scope for Plantation Crops

Expansion in Areas

These crops are grown in tropical regions only in a limited area. The area under these crops can be extended under the non-traditional areas where land is fertile and availability of irrigation facilities. An increase in area and application of irrigation through drip irrigation can increase area and production.

Export Potential

Plantation crops have huge potential for export of the main product and sub-products/byproducts (coir, cocopeat). The products of plantation crops have internal demand in several industries. The total earnings from the export of plantation crops produce is about 27 per cent of agricultural produce and it accounts for about 4.8 per cent of total exports. The export earnings from by-products (coir and coir products) are about Rs 260 million per annum.

Employment Generation

The labour requirement in plantation crops is throughout the year. The labour is required on the farm and off the farm *i.e.* in factories or processing units. In Kerala, coconut crops provide employment to 70 million-man days per year.

Diversification of Crops

The plantation crops can be adopted to increase diversification or avoid monoculture. The diversification increases the area under sustainable agriculture as well as reduces the risk in crop production.

Technologies to Meet our Yield Gap

The crop wise institutes are available to supply the latest technology to the farmers/growers. These institutes are available in the crop growing areas, which are helpful to increase area, production, productivity and quality of the produce. These institutes also provide all kinds of guidance/help to the farmers regarding production technology, processing and export of produce.

- ICAR institutes (NRCs, Directorates).
- State Agriculture Universities.
- State Government Institutes.
- Research Stations/Regional Research Stations.

Economic Importance

These crops provide economic benefits to the farmers in terms of better returns per unit area and good market at the national as well as international levels.

Export Earnings

Earnings from the export of main products and byproducts of plantation crops are higher than those from other crops. Plantation crops cover less than 3 per cent of the area of cultivable land (4.29 m ha out of the total area under crops, *i.e.* 143 m ha), but the export earning is about Rs 16000 million, which is about 12.72 per cent of total exports and 75 per cent of earnings from total agriculture exports.

Position in the World

As per statistics, India is the leading country in the production of certain plantation crops at the global level. Production share is given below:

- India ranks 1st in global production of arecanut with 56.6 per cent of world production.
- India ranks 1st in coconut production (30.9 %) with the production of 20309 million nuts and 2nd in productivity (*i.e.* 9346 nuts/ha) after Vietnam.
- India ranks 2nd in global production of cashewnut with 13.2 per cent of the world production.
- India ranks 4th in rubber production (6.8 %), 2nd in tea production (21.2 %), 8th in coffee production (3.2 %) of world production.

Employment Opportunities

It provides employment to the labour throughout the year directly on farms or in processing units/industries. The number of labour requirements or employment generation varies from crop to crop *e.g.*

- Tobacco provides employment to 20 lakh people.
- Coconut provides employment to 10 lakh people.
- Cashewnut provides employment to 5 lakh people.

Industrial Importance

The plantation crops provide raw materials to industries like Cottage and Micro Small & Medium Enterprises (MSME). The coconut provides fibre from coconut husk, which is about 2.8 lakh MT annually to different industries.

Conserving Soil and Ecosystem

- These crops are helpful in protecting the land from wind and soil erosion and conserving the ecosystem. These crops are planted on hilly slopes and protect the land from sliding.
- Some crops like cashew can be grown easily on barren land, which are helpful to protect soil erosion, conserve moisture and increase the value of barren land.

Institutes Working on Plantation and Beverage Crops

- *Coconut*: Central Plantation Crop Research Institute, Kasargod, Kerala.

  Coconut Development Board, Kochi, Kerala.
- *Arecanut*: CPCRI, Regional Research Station, Vittal, Karnataka.

  Directorate of Arecanut and Spices Development, Kozhikode, Kerala.
- *Cocoa*: CPCRI, Regional Research Station, Vittal, Karnataka.
- *Rubber*: Rubber Research Institute of India, Kottayam, Kerala.

  Rubber Board, Kottayam, Kerala.
- *Cashew*: Directorate of Cashew Research, Puttur, Karnataka.

  The Cashew Export Promotion Council of India (CEPC), Mundakkal, Kollam city, Kerala.
- *Oil Palm*: National Research Centre, Elur, Pedavegi, Andhra Pradesh.
- *Palmyrah*: Palm Research Station, Srivaliputtur, Tamil Nadu.
- *Tea*: Tea Research Institute, United Planters Association of South India (UPASI), Valparai, Tamil Nadu.

  Tea Board, Kolkatta, West Bengal.
- *Coffee*: Central Coffee Research Institute, Balehonnur, Karnataka.

  Coffee Board, Banglore, Karnataka.

Chapter 2

High-Density Planting

High-density planting (HDP) was first used in Europe at the end of 1960. In India, HDP has been proven useful in many fruit crops, *e.g.* pineapple, banana, mango, apple and citrus.

Growing more number of plants per unit area without adverse effects on yield and quality of fruits.

In other words, planting of more number of plants than optimum, through manipulation of tree size.

Aims of High-Density Planting (HDP)

- Light interception (absorbed more photosynthetically active radiation by the canopy).
- Maximize the production of assimilates and its conversion into fruits of acceptable quality.
- Optimization of light interception and productivity over the life of the orchard.
- To enhance the productivity of fruit crops.
- To maintain a balance between vegetative and reproductive load without impairing the plant health.

Principles of High-Density Planting

- Higher and early yields with best-quality fruits through planting more number of plants per unit area.
- Maximum possible returns per unit of inputs through efficient exploitation of natural resources. Best use of vertical and horizontal space per unit area and per unit time.

- ☆ The balance between vegetative growth and fruiting means growing fruit, not trees.
- ☆ Increase fruiting branches and decrease structural branches.
- ☆ Training and pruning in such a way that each branch has minimum shade effect on others.

Advantages

- ☆ It induces precocity, increases yield and improves fruit quality.
- ☆ Reduces cost of production by minimizing labour cost.
- ☆ Support mechanization in the production of fruit crops.
- ☆ Increase fertilizers, water, solar radiations, fungicides, weedicides and pesticide efficiency.

Disadvantages

- ☆ In the long run, it results in heavy competition for space, nutrients, sunlight and water.
- ☆ Overcrowded growth of the canopy increased humidity and reduced cross ventilation of air in the orchard, which are conducive to more incidence of pests and diseases.
- ☆ The trees in close planting become crowded very soon and decline in yield.
- ☆ Efforts to breed dwarfing rootstocks to control the growth of mature trees have largely been unsuccessful.
- ☆ Effective canopy management is the largest barrier to the success of high-density orchards.
- ☆ The growth regulators used for restriction of growth in high-density orchards were reported to have some persistence in the harvested fruit and soil.
- ☆ Reduce fruit size and quality.

Type of High-Density Planting Systems

Different HDP systems have been tried mostly in apples and peaches and to a lesser extent, in citrus.

i) **Hedge row system**: Decrease the spacing within a row to form closed hedge rows and each row is bordered by two alleys for the movement of the tractor.

ii) **Multi-row system or bed system**: Multi-rows consist of 2,3,4,5, or 6-row beds separated by alley rows. The loss of space under alleyways in a hedge row system could be alleviated either by reducing the number of alleys in a multi-row or bed system.

iii) **Full-field system**: In the entire field orchard, the rows are so close together that the tractors cannot pass along the rows and management should, therefore, be carried out with over-row tractors.

iv) **Meadow orchard system**: The area is completely covered by plants using dwarfing rootstock (apple(70000 trees/ha) and growth regulators *e.g.* SADH and Alar (sprays @ 2000 to 2500 ppm).

v) **Fruit wall or narrow hedge row system**: Planting very close like a wall in a row.

vi) **Tatura trellis system**: Plants are trained in 'V' shape and planted close together in a single row. Each tree has two limbs trained at 60° from the horizontal across the inter-row alleyway with the help of training wires (*e.g.* peach and pear).

Concept of HDP

A) Light interception

The relation between light interception and productivity forms the basis of high-density planting. In the apple orchard, yield increased linearly as light interception was increased up to about 60 per cent of available light. Light interception 81 per cent and 83 per cent have been reported in very closed-spaced apple orchards. An interception of 30 per cent of annual radiation is required to maintain good production efficiency in citrus. Factors affecting light interception are as under:

i) **Canopy:** A shallow canopy up to 1.5 to 2.0 m in depth is needed to achieve maximum efficiency in trapping sun energy and converting the highest proportion of metabolism into quality fruits. The effect of spacing on a light interception in citrus orchards is as under:

   46 per cent of the ground area is covered at 30 feet x 30 feet spacing.

   64 per cent of the ground area is covered at 20 feet x 20 feet spacing.

ii) **Rectangularity:** A square system intercepts more sunlight and gives more productivity than a rectangular one.

iii) **Row orientation:** North-South row orientation in a hedge row gives the best light exposure on both sides of the hedge row and gives more yield.

iv) **Canopy bearing volume and productivity:** The volume of the outer 3 feet or one metre of tree canopy is termed as canopy bearing volume. The yield potential is directly and linearly related to canopy bearing volume.

v) **Production efficiency**: Defined as yield per unit of production capacity. Production efficiency may be calculated as yield per unit of ground area, canopy area, canopy volume or canopy bearing volume.

vi) **Optimum bearing volume and productivity over the life of orchard:** High-density planting orchards are designed to maximize the average bearing volume, yield and net return over the life of the orchard.

vii) **Induction of precocity**: An induction of precocity using dwarfing and semi-dwarfing rootstocks is an essential feature of all high-density plantings. Traditional pomological techniques, such as (summer pruning, tying down of branches, root pruning and trunk girdling) and growth regulators are used to induce precocity where dwarfing rootstocks are not available. Chlorocholine chloride (CCC) has proved more useful in inducing precocity on pears, while succinic acid-2-2-dimethylhydrazide (SADH) on plums.

B) Utilization of Resources

i) **Production capacity of soil:** The production capacity of soil decreases with an increase in plant density due to nutrient competition. The productivity increases due to a larger number of trees per unit area, but production per plant decreases. High-density planting has higher water requirements, especially during the early years of establishment of orchard.

ii) **Soil and leaf nutrients:** Nutrient removal is more in high-density planting. The tree density did not affect the nutrient composition of the leaves.

C) Planting density on tree growth, yield and fruit quality

i) **Tree growth:** Tree growth expressed as trunk cross-sectional area, showed an inverse relationship to planting density. Shading reduced total growth increment. The number of roots per tree decreases with an increase in tree density.

ii) **Yield:** With the increase in density, fruit weight and yield per plant decrease while the yield per unit area increases.

iii) **Fruit quality**: Shading affects the fruit quality. The anthocyanin content in the skin of apple fruits increases with light intensity up to 50 per cent of full sunlight, while the shaded area produces few small and green fruits as they receive less (35 per cent) sunlight. The total soluble solids (TSS) content in the fruits is positively correlated with the interception of sunlight.

D) Effect of planting density on disease and pest incidence

The application of fungicides and pesticides is more efficient in high-density planting due to better spray interception. However, disease and pests were observed more in high-density planting than in normal spacing due to poor aeration and more humidity.

Components of High-Density Planting to Control Tree Size

1. Genetic control

    i. Dwarf scion/ cultivars

    ii. Dwarf rootstocks or interstock

2. Mechanical control (training and pruning)
3. Cultural methods
4. Growth retardants/ Chemicals
5. Incompatible rootstock
6. Viral dwarfing

1. Genetic control

i) Dwarf scion/ cultivars: Some scions are dwarf genetically or have dwarfing effects on specific rootstocks.

Table 2.1: Genetically Dwarf Scion Varieties

Fruit crops	*Varieties*	*Desirable characteristics*
Apple	Red Chief, Oregon Spur, Red Spur, Stark Spur Gold, Well Spur, Star Crimson Delicious.	Bear more spurs, dwarf, high yield.
Banana	Dwarf Cavendish (AAA), Pusa Dwarf.	Dwarf stature, high yield.
Cherry	Compact Lambert, Meteor, North Star.	High-yielding, self-fruitful, dwarf.
Guava	Pant Prabhat.	Less spreading, high yielder.
Mango	Amrapali, Arunika, Mallika.	Precocious, regular bearer.
Litchi	Calcuttia, China.	Upright growth habit.
Papaya	Pusa Nanha, Pusa Dwarf, Red Lady 786 .	Dwarf, tends to bear at lower heights.
Peach	Red Haven, Candor.	Dwarfing, high-yielding.
Sapota	PKM 1, PKM 3.	Columnar shape, dwarf stature.

ii) Dwarf rootstocks or interstock: Rootstocks have a profound effect on tree vigour, precocity, fruiting, fruit quality and longevity of orchards in different fruit crops.

Table 2.2: Genetically Dwarf Rootstock Varieties

Fruit crops	*Dwarfing rootstocks*
Apple	M 27, M 9, T 337, M 26, Bud 9, P 22, Ottawa 3, MM 111, MM 106, Jork 9, Bemali.
Ber	*Ziziphus rotundifolia, Z. nummularia.*
Citrus	Trifoliate orange (*Poncirus trifoliata* var. Flying Dragon), Troyer citrange, Citrangequat, Feronia, *Severinia buxifolia*, Karna Khatta, Sour orange.
Guava	*Psidium friedrichsthalianum, P. pumilum, P. chinensis*, Pusa Srijan or Aneuploid No. 82.
Mango	Vellaikulumban for Alphonso, Olour for Himsagar, Langra.
Pear	Quince C
Peach	Siberian C, St Julien X, *Prunus besseyi*, Rubira.
Plum	Pixy.
Cherry	Colt, Charger.

- ☆ Interstock for mango: Kalapadi interstock for Langra, and Anupam interstock for Amrapali.
- ☆ Commercial cultivars of pear like Bartlett, Bosc, Eldorado need 'Old Home' as interstock

2. Mechanical control (training and pruning)

It includes hedging, topping, pruning and tree thinning.

i **Hedging and topping:** In citrus, removal of apical dominance stimulates lateral bud break.

ii) **Summer pruning:** In apple, it is used to maintain tree productivity and tree size. During summer pruning, tying down of branches, root pruning and trunk girdling can be used to induce precocity.

iii) **Hand pruning:** Very little hand pruning is practised because of higher cost and greater tree vigour.

iv) **Tree thinning:** Remove trees or alternate rows to avoid over-crowding.

3. Cultural methods

Cultural methods such as regulated supply of irrigation, nitrogen, root restriction, spacing arrangement, poorly adapted rootstock and budding can be used to a limited extent under special circumstances to control vegetative growth under high-density planting.

4. Growth retardants/ chemicals

Succinic acid-2,2-dimethyl hydrazide (SADH) gave good results in plums, while cycocel proved more useful to retard growth and bringing young, vigorous pear trees into the crop.

5. Incompetable rootstock

Plant size decreases due to an insufficient supply of nutrients and water by rootstock e.g. pear on quince.

6. Viral dwarfing

Strains of exocortis virus that have a dwarfing influence are being used in Australia for intentional inoculation during citrus buddings on Rangpur lime rootstocks to reduce tree size.

Recommended planting density for different fruit crops

The development of a series of dwarfing and semi-dwarfing rootstocks for each fruit tree species have tremendous practical value in the commercialization of high-density planting.

- ☆ Amrapali mango at 2.5 m x 2.5 m (1600 plants/ha) spacing yields 24.46 t/ha.

- ☆ Dashehari at 2.5 m x 3.0 m (1333 plants/ha) spacing yields 18.6 t/ha.
- ☆ Dashehari at 12 m x 12 m (69 plants/ha) spacing yields 1.6 t/ha.
- ☆ Kinnow on Troyer citrange can be planted at spacing 1.8 m x 1.8 m (3086 plants/ha).
- ☆ Kinnow on Karna Khatta can be planted at a spacing of 3 m x 3 m (1111 plants/ha).
- ☆ Banana (Robusta) can be planted at spacings 1.2 m x 1.2 m, 1.5 m x 1.5 m, 1.8 m x 1.8 m, 2.1 m x 2.1 m.
- ☆ Pineapple can be planted at 60 cm x 90 cm x 120 cm (15000 plants per ha). 30 cm x 60 cm x 120 cm (35000 plants/ha).

Chapter 3

Use of Rootstocks

Rootstocks

Once the plant is transplanted, then rootstocks cannot be changed throughout the life of the plant, so attention should be paid to all points while selecting the rootstocks. The functions of the rootstocks should be for an extended period. Hence, before choosing a rootstock, it is essential to assess the proposed site in terms of soil and climatic factors because rootstock performance is affected by these factors. Classification of rootstock on the basis of propagation methods: Seedling and clonal.

i) Seedling Rootstocks

- Simple, economical and multiply easily.
- Well adapted to mass propagation.
- Rootstock tends to grow deeper.

  Drawback: Genetic variability.

ii) Vegetative/Clonal Rootstocks

- Different vegetative methods of propagation: cuttings, layering, stooling or tissue culture are used to prepare vegetative rootstocks.
- Vegetatively propagated rootstocks have some genetic makeup. These rootstocks perform the same growth behaviour and response on the scion. In case of polyembryonic crops/cultivars, the nucellar seedling is of clonal origin and true to type to the mother plant.
- The rootstocks resistant/tolerant to biotic or abiotic stress are propagated through conal propagation. The most commonly used clonal rootstocks are prepared in grapes, apple, pear, plum, citrus and cherry.

Characteristics of Rootstocks

1. It should be multiplied easily by seed or vegetative propagation.
2. It should be available easily.
3. It should provide enough strength or support to the plant.
4. It should produce an upright and smooth stem for easy propagation.
5. It should survive easily under a wide range of soils and climatic conditions.
6. It should be compatible with scion cultivars.
7. It should have a positive influence on scion.
8. It should tolerate biotic (insect-pest and diseases) and abiotic (low and high temperature, moisture) stress.
9. It should give a dwarfing effect on the scion and favour heavy bearing.

Beneficial Effects of Rootstocks

Rootstocks have pronounced effects on scion cultivars by influencing production efficiency, yield, quality of fruits, adaptability, tree vigour, resistance to biotic and abiotic stresses *etc.* Rootstocks not only provide physical support, *i.e.* anchorage by penetrating the roots deep into the soil, but also regulate the uptake of nutrients and moisture from the soil. There are many plant cultivars which have been selected for their desirable fruit or ornamental qualities, but they do not possess a good root system and thus require grafting on another genotype to give a satisfactory plant.

1. Tree Vigour

a) Apple rootstocks

i) **Very dwarfing:** M 27 is a cross between M 9 and M 23 produced in 1929 from East Malling Research Station. This can be planted as close as 0.5 m in a row and 1.5 m between rows (M: Malling, MM: Malling Merton).

ii) **Dwarfing:** M 9 is a chance seedling of apple widely used as dwarfing rootstock.

iii) **Semi dwarfing:** M 7, M 29, MM 106.

iv) **Vigorous:** MM 104, MM 109, MM 111.

v) **Very vigorous:** M 16 and M 25.

a) Mango rootstocks

Dwarfing rootstocks of mango are Olour, Taimuria, Rumani, Nekkare, Kurukkan, Vellaikulumban and Totapuri Red Small.

c) Citrus rootstocks

Flying Dragon is a dwarfing rootstock used for high-density planting in different parts of the world.

2. Anchorage

- It is the important and primary function of the rootstock to provide support/strength to the plant with a deep root system.
- The most dwarfing rootstocks like M 27 and M 9, usually require staking, particularly in the early years due to the brittle nature of their roots.
- Citrus rootstocks are grouped into shallow, medium and deep-rooted according to their penetration depth in the soil.

  Deep-rooted: Rangpur lime. Rough lemon and Karna Khatta.

  Medium rooted: Cleopatra mandarin, Sweet lime and Sweet orange.

  Shallow rooted: Trifoliate orange and its hybrids like Troyer citrange and Carrizo citrange.

3. Leaf Nutrient Status

- The leaf nutrient status of the cultivar is also governed by the rootstocks.
- The uptake of potassium is lower in sweet orange on all the rootstocks except on Rangpur lime and Mosambi seedlings.
- The petiole nitrogen contents have been reported higher in Anab-e-Shahi on Dogridge rootstock, whereas the highest P and Na content were accumulated by Anab-e-Shahi on its own roots, but the highest accumulation of K was found on St. George rootstock.

4. Fruit Characters

- Rootstocks influence fruit characteristics like fruit weight, rind thickness, seed number, number of carpels *etc.*
- In Croog mandarin, the highest fruit weight was obtained on Rough Lemon and the least on Trifoliate orange rootstock.

5. Precocity in Bearing

- The dwarfing effect of rootstocks brings precocity in bearing.
- M 9 and M 27 rootstock of apple induce precocity in bearing.

6. Staggering Fruit Maturity

- Often, the horticultural produce faces a glut when it arrives in the market, and the farmers are compelled to make distress sales due to the perishable nature of horticultural produce.
- The Kinnow fruit on Troyer citrange is ready to harvest by the end of November, which is considered to be very early, whereas the maturity may be staggered up to mid-January by using two other rootstocks, *viz.,* Karna Khatta and Soh Sarkar. Thus, the availability of Kinnow fruits can be extended invariably by the use of rootstocks, which are not only useful to the farmers but consumers as well.

7. Flowering, Fruiting and Yield

In most of the fruit crops, rootstocks positively influence flowering, fruiting and yield. However, yield per tree remains less in dwarfing trees, but yield per unit area or productivity per unit canopy area remains higher. The dwarfing effect helps to produce early and profuse flowering on scion as compared to vigorous rootstocks, *e.g.* apple rootstocks.

8. Fruit Size

- ☆ Fruit size and yield varies with the kind of rootstocks, *e.g.* dwarfing roostocks induce precocity in bearing but reduce size of fruits.
- ☆ M 9 and M 27 dwarfing rootstocks produced small sized fruits than seedling rootstocks. In guava, the fruits of Allahabad Safeda are smaller in size on Aneuploid No. 82 than on its roots.

9. Fruit Quality

- ☆ The rootstocks are known to regulate the fruit quality directly or indirectly *e.g.* the fruits on the dwarfing guava rootstocks *Psidium pumilum* had higher total soluble solids and total sugars.
- ☆ Fruits of Allahabad Safeda had higher total soluble solids (TSS) on Aneuploid No. 82 than on Allahabad Safeda. Kinnow fruits have more TSS on Troyer citrange than Karna Khatta and Soh Sarkar rootstocks.

10. Soil Conditions

Annona glabra rootstock for Annona is resistant to wet and damp soil conditions.

11. Effect on Biotic Factors (Insect-pest, diseases and nematode)

- ☆ On the basis of response to insect-pests, rootstocks are classified as resistant, tolerant and susceptible.

i) Insect Pest

- ☆ Jamun rootstock (*Syzygium densiflora*) is resistant to termite attack.
- ☆ Malling Merton (MM) series and Northern Spy rootstocks of apple are resistant to woolly apple aphids.

ii) Diseases

Cleopatra mandarin rootstock is resistant to viral diseases. Dog Ridge and 1613 rootstocks of grapes are resistant to phylloxera root louse.

iii) Nematode

Rootstocks of almond Almen, Almen 88, Almen 201 are nematodes resistant and fig rootstock (*Ficus glomerata*) is resistant to nematodes.

Graft Incompatibility

When stock and scion are unable to join due to physical, anatomical, physiological, biochemical, nutritional deficiency, viruses and diseases. Graft incompatibility is mainly classified into two groups:

i) **Translocated:** The substances transported from stock to scion or vice versa act as toxins.

ii) **Non-translocated (Localized):** Poor union/ physical contact due to starch accumulation and necrosis.

Overcome Graft Incompatibility

Graft incompatibility decreases the life of plants and sometimes plants fail to establish. Graft incompatibility is encountered in many fruit species such as ber, pear, walnut, citrus and cherry. However, at the same time, graft incompatibility is used as a tool to give a dwarfing effect on the plant.

i. Use compatible rootstocks and scion.

ii. Select insect-pest, disease and virus-free material.

iii. Follow standardized techniques and proper time for propagation.

iv. Use inter-stock to overcome incompatibility.

v. Provide proper rooting media/ soil and climatic conditions.

Production Technologies for Cultivation of Major Fruits

Chapter 4

Mango (*Mangifera indica* L.): The King of Fruits

Family: *Anacardiaceae; 2n=40*

Introduction

- Mango has been in cultivation in the Indian sub-continent for over 4000 years ago (De Candolle,1904).
- Hiuen Tsang was the first foreigner to bring Mango to the notice of the outside world.
- The greatest tribute to this esteemed plant was paid by emperor Akbar, who established the Lakh Bagh (a mango orchard with 1,00,000 plants) in Darbhanga, Bihar in 1580.
- Mango planting was done at Shalimar Bagh (Lahore) and Mughal Garden, Pinjore.
- Ain-e-Akbari also has information about mango cultivation.
- Mango is a good source of vitamin 'A' and contains about 4800 IU/ 100g pulp.
- India ranks 1st in production (39.4 %) of mango.
- Mango is exported in large quantities. Mango export from India is 49,180 MT, amounting to Rs 38,234 lakh.
- Mango is also called a bathroom fruit.
- Mango pulp is important among the processed fruit products exported.

- In India, major mango-producing states are Uttar Pradesh (24.4 %), Andhra Pradesh (22.7 %), Bihar, Karnataka, Gujarat *etc.*
- In Haryana, major mango-producing districts are Yamunanagar, Ambala, Karnal, Panchkula, Kurukshetra and Panipat.

Area and Production of Mango in India

- Area: 23,50,300 ha.
- Production: 207,72,300 MT.
- Productivity: 8.84 MT/ha.
- Per cent of total fruit area: 33.3.
- Per cent of total fruit production: 19.3.
- Highest productivity in Punjab (17.56 MT/ha).

Origin

Indo Burma region extends from Eastern India and Southern China across Southern Asia.

Soil and Climate

- It grows on a wide range of well-drained soil, from alluvial to lateritic.
- Deep (2.0 to 2.5 m) soils without a hard pan are the best.
- Mango is sensitive to saline conditions. The favourable pH for growing mango is 5.5 to 7.0 (Whiley, 1984).
- Black cotton soils are unsuitable for mango cultivation.
- The mango orchard should be at least 2 km away from brick-kilns.
- Mango is basically a tropical fruit according to its origin but performs well in subtropical conditions. Hence, it is termed as a pan-tropical crop.
- It can be grown up to 1100 to 1400 m above the mean sea level (msl).
- High humidity, rainfall and frost during flowering are harmful.
- The best temperature for mango crops is 24 to 27°C.
- Moderate temperature during winter in South India allows early flowering (between September to November) and result in two crops in a year under Kanyakumari conditions.

Cultivars

- In India, about 1,500 varieties of mango are grown including 1,000 commercial varieties. Almost all of these are of monoembryonic type.
- About a dozen polyembryonic varieties also exist in India *viz.* Bappakai, Chandrakaran, Goa, Kurukkan, Olour, Bellary, Kasargod, Mazagaon, Nileshwar Dwarf and Salem.
- Based on the time of maturity mango varieties are be classified as under:
  - Early maturing: Bombay, Bombay Green, Himsagar, Kesar, Suvarnarekha.

- Mid-season maturing: Alphonso, Mankurad, Bangalora, Vanraj, Banganapalli, Dashehari, Langra, Kishen Bhog, Zardalu, Mankurad.
- Late maturing: Fazli, Fernandin, Mulgoa, Neelum, Chausa.

Table 4.1: Commercial Cultivars of India

Sr.No.	*Cultivar*	*Region*	*Characteristics*
1.	Dashehari	North	Mid-season maturing, biennial bearer and suitable for canning.
2.	Langra	North	Mid-season maturing, turpentine flavour and most prone to fruit drop.
3.	Chausa	North	Late maturing, sweetest variety and highly susceptible to malformation.
4.	Bombay Green	North	Earliest maturing and highly susceptible to malformation.
5.	Fazli	East	Late maturing variety and with a large sized fruits.
6.	Himsagar	East	Regular bearer.
7.	Kishenbhog	East	Mid-season maturing.
8.	Gulabkhas	East	Rose flavour and reddish blush towards the base.
9.	Alphonso	West	Finest export quality cultivar, regular bearer and highly susceptible to spongy tissue.
10.	Kesar	West	Good processing quality.
11.	Pairi	West	Regular and heavy bearer.
12.	Mankurad	West	Mid-season maturing.
13.	Neelum	South	Regular and heavy bearer, late season maturing.
14.	Banganapalli or Safeda	South	Early maturing.
15.	Suvarnarekha	South	Early bearer and attractive pinkish-red fruits.
16.	Totapuri	South	Regular and heavy bearer.
17.	Mulgoa	South	It is an excellent cultivar for late season in the South.
18.	Pusa Surya	North	Regular bearer and shelf life is longer (12 days after ripening).
19.	Arka Neelachal Kesari	South	Extra early and free from fruit flies.

Table 4.2: Mango Hybrids

Sr.No.	Centre	Hybrid	Parents	Characteristics
1.	IARI	Amrapali	Dashehari x Neelum	Dwarf and regular bearer.
2.	IARI	Mallika	Neelum x Dashehari	Highest vit. A and regular bearer.
3.	IARI	Pusa Lalima	Dashehari x Sensation	Regular bearer and fruits have red peel.
4.	IARI	Pusa Shrestha	Amrapali x Sensation	Regular bearer, fruits have red peel and orange pulp.
5.	IARI	Pusa Peetambar	Amrapali x Lal Sundari	Regular bearer and good sugar: acid blend.
6.	IARI	Saisugartha	Totapuri x Kesar	Regular bearer and free from malformation.
7.	IARI	Pusa Arunima	Amrapali x Sensation	Regular bearer with export quality fruits.
8.	IARI	Pusa Pratibha	Dashehari x Amrapali	Regular bearer and red peel colour on golden yellow background.
9.	IIHR	Arka Aruna	Banganapalli x Alphonso	Dwarf, early maturing, regular bearer, fibreless and free from spongy tissue.
10.	IIHR	Arka Anmol	Alphonso x Janardhan Pasand	Semi-vigorous and regular bearer.
11.	IIHR	Arka Neelkiran	Alphonso x Neelum	Late maturing with export quality fruits.
12.	IIHR	Arka Puneet	Alphonso x Banganapalli	Free from spongy tissue.
13.	CISH	Ambika (CISH-M-1)	Amrapali x Janardhan Pasand	Regular bearer and have export potential.
14.	CISH	CISH-M-2	Dashehari x Chausa	Late maturing, firm flesh and scanty fibres.
15.	CISH	Arunika	Amrapali x Vanraj	Regular bearer.
16.	RFRS, Sangareddy	Manjeera	Rumani x Neelum	Early maturing and regular bearer.
17.	BAC, Sabour	Prabha Sankar	Bombay x Kalapady	Regular bearer.
18.	RFRS, Vengurle	Ratna	Neelum x Alphonso	Early maturing, regular bearer and free from spongy tissue.

19.	RFRS, Vengurle	Sindhu	Ratna x Alphonso	Seedless and high-yielding.
20.	IARI	Pusa Deepshikha	Amrapali x Sensation	Red peel, orange colour pulp and excellent keeping quality.
21.	IARI	Pusa Manohari	Amrapali x Lal Sunderi	Reddish shoulder and tolerant to mango malformation.
22.	IIHR	Arka Suprabhat	Arka Anmol x Amrapali	Pulp is solid with a deep orange colour and shelf life at room temperature is 8 to 10 days.

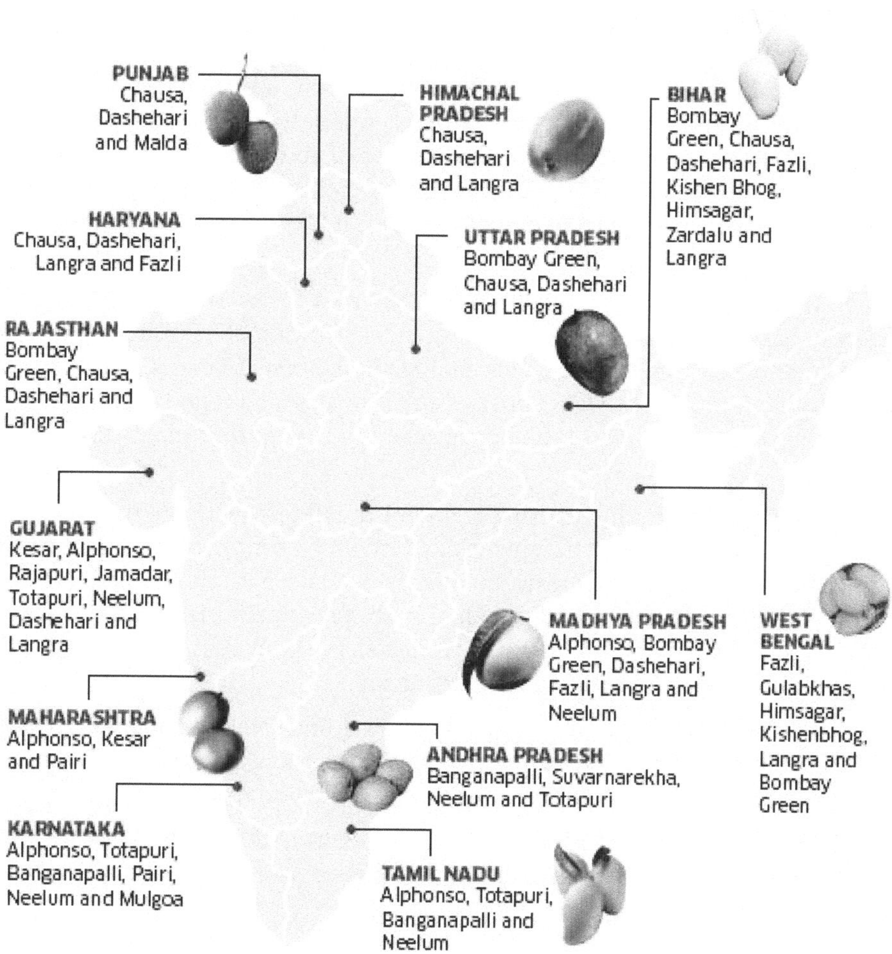

(Source: APEDA)

Fig. 4.1: Map of Commercial Varieties Grown in India

Propagation Methods

- ☆ **Sexual**: By seed
- ☆ **Asexual**
  1. Veneer grafting
  2. Epicotyl or stone grafting
  3. Inarching or Approach grafting
  4. Cleft grafting
  5. Budding (Patch, Shield and Forekert)
  6. Stem cutting

1. Veneer grafting: It is similar to side grafting except that only one side of the scion is cut in a slanting manner. Mango is commercially propagated by this method in North India.

2. Epicotyl or Stone grafting: The stone remains attached to the plant at the time of grafting during July-August. The two-week-old seedling is headed back and the vertical split is given in the centre. The scion of about two months old seedling is prepared in a wedge shape. The scion is then insert into the splitted rootstock and the union is tied with a polythene strip. This method is commercially followed in the mango in Konkan region.

3. Inarching or Approach grafting: The rootstocks are grown in pots and brought closer to the mother trees. A 5 to 10 cm portion of bark along with some wood is removed from the rootstock at a height of about 15 cm. A similar process is repeated on the scion. The exposed cambium is joined together and tied with polythene strips. The scion remains attached to the mother plant till the union takes place.

4. Cleft grafting: Clift grafting is one of the oldest methods of propagation and is most commonly used in topworking (crown grafting) or grafting on scaffold branches. It is done during the dormant season. Make a vertical split cut of 5 to 8 cm in the centre of the scaffold branch or stub. Take an 8 to 10 cm long scion and cut its basal portion of about 5 cm in wedge shape (v shape) and insert it in the stock. Wrap the union portion with grafting tape.

5. Budding (Patch, Shield or Forket): Budding can be done by inserting or putting a slightly swollen bud of a similar size to the patch made on the stock. Cover the budded portion with budding tape, keeping the bud/eyes open.

6. Stem cutting: Take 15-20 cm long cutting from the plant having more C: N ratio. Cutting should be collected from lateral shoots for better rooting and it can be treated with rooting hormone for early and better rooting.

Rootstocks and Scion

Rootstock: The part of the grafted plant below the bud union, used as rooting and supporting the plant is known as the rootstock.

Dwarfing rootstocks: Creeping, Olour, Rumani, Totapuri Red Small.

Vellaikulumban: Dwarfing and allopolyploid.

Kurukkan: Polyembryonic and salt tolerant.

Scion: The part of a grafted plant that develops into a shoot and bears leaves, flowers and fruit is called a scion.

Rootstock Influence

1. Dwarfing Effect: Following rootstocks gives the dwarfing effect on scion:

   Rumani on Dashehari; Olour on Himsagar and Langra; Vellaikulumban on Alphonso.
2. Salt resistant: Kurukkan.
3. Salt tolerant: 13-1, Moovandan, Nekkare, Anna and Gomera.
4. Polyembryonic: Pahutan, Olour on Baneshan scion; Pahutan, Olou, Bappakai on Neelum scion and Saber on Haden scion.
5. Interstock: Kalapadi interstock for Langra, and Anupam rootstock for Amrapali.

Planting

- ✰ Land preparation.
- ✰ Generally, a square system of planting is followed in 8 m x 8 m and 9 m x 9 m spacing.
- ✰ Spacing followed for high-density planting is 2.5 m x 2.5 m (1600 plants/ha).
- ✰ Dig the pit of 1 m x 1 m x 1 m and keeps it open for about 1 to 2 months.
- ✰ Fill the pit with a mixture of ½ topsoil + 50 kg farmyard manure + 2 kg single super phosphate (SSP) + 50 ml chlorpyriphos.
- ✰ The best time for planting is July to September and February to March.
- ✰ During planting, the earth ball should remain intact with roots and the graft union should remain about 15 cm above the ground level.

High-Density Planting (HDP)

- ✰ The majority of existing orchards are conventional and planted at 10 m x 10 m spacing.
- ✰ For HDP of Amrapali, 2.5 m x 2.5 m (1600 plants/ha) spacing is recommended.
- ✰ For HDP of Dashehari, 4 m x 4 m spacing is recommended (625 plants/ha).
- ✰ For high-density planting, 6 m x 4 m, 5 m x 5 m, 5 m x 3 m and 6 m x 2 m spacings is adopted in new plantation.
- ✰ Spacing for medium-density plantation is 4.5 m x 4.5 m.

- Spacing for ultra-high-density plantation (UHDP) is 3 m x 1 m or 3 m x 2 m.

Varieties for High-Density Planting

Amrapali, Totapuri, Imam Pasand, Baneshan, Alphonso, Ratna, Neelum, Mallika, Himsagar, Dashehari. Regular bearer varieties perform better under high-density planting.

Orchard Floor Management/Weeding

- **Mechanical method:** Hand weeding, mulching, hoeing *etc.*
- **Chemical method:** Apply herbicides.
- **Mulching**: Organic mulch (paddy straw) or inorganic mulch (black polythene).
- **Inter cropping:** Legume crops, turmeric, ginger or short-duration fruit trees.

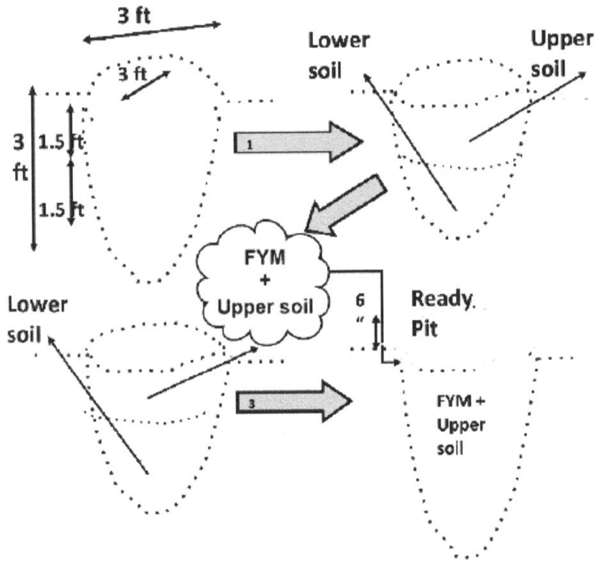

Fig. 4.2: Digging and Filling of Pit

Manures and Fertilizers

Add 100 kg farmyard manure (FYM), 500 g urea, 1.0 kg SSP and 1.0 kg sulphate of potash (SOP).

- Apply FYM and SSP during December.
- Apply urea and sulphate of potash (SOP) during February.
- Apply an additional dose of urea in June during on year of fruiting.

Irrigation

- ☆ First irrigation should be given just after planting. Young trees must be frequently irrigated until the establishment of the plant.
- ☆ More water is required during summer to promote proper vegetative growth.
- ☆ Irrigation must be avoided during flowering.
- ☆ Sufficient moisture should be maintained during the fruit development stage.
- ☆ However, the quantity and frequency of irrigation may be modified on the basis of soil type and climatic conditions.

Training and Pruning

- ☆ Training of plants is essential during the initial 2-3 years.
- ☆ In mature trees, pruning is generally done to remove the diseased and criss-crossed branches.
- ☆ Pruning results in the removal of additional growth flushes increases productivity and stimulates precocity.
- ☆ Pruning should be done after the harvest of the crop, *i.e.* during the 3rd week of July.
- ☆ Tree growth and flowering are regulated through a combination of regular pruning and application of paclobutrazol @ 10 g a.i./plant in September.
- ☆ Advantages – Reduced gestation period, increased yield per unit area and excellent fruit quality.

Growth Regulators

1. A spray of ethrel (250 or 500 ppm) in the month of December increase floral branches and yield.
2. Application of GA_3 @ 50 ppm (gibberellic acid) increases the number of perfect flowers as well as the viability of pollens.
3. Application of NAA @ 20 ppm (1-naphthalene acetic acid) at flowering increases fruit retention.
4. Application of 2,4-D @ 20 ppm after fruit set reduces fruit drop.
5. During February, 0.5 per cent urea or 1.0 per cent potassium nitrate may be sprayed to induce flowering if trees do not flower by that time.
6. Post-harvest treatment with GA_3 @ 200 ppm has been found to be highly effective to retard ripening.
7. Application of paclobutrazol @ 10 g a.i. on non-bearing trees during the first fortnight of September induces flowering, fruit set and yield during off years.

Harvesting

- ☆ Slight colour development on the shoulder.

- When one or two ripe fruits fall from the plant individually (*tapka* stage).
- When the specific gravity of fruits reaches between 1.01 and 1.02.
- Mango is a climacteric fruit. So, harvesting the fruit at the physiological maturity stage.
- Fruits mature between 90 and 120 days after fruit set.
- Harvesting is done by plucking fruits manually or by mechanical pluckers.
- The average fruit yield is 8 MT/ha.

Ripening

- Ripening starts 5 days after harvesting (Ethylene @ 100 ppm).
- Mango is generally stored at 8 to 9°C temperature and 85 to 90 per cent relative humidity for 2 to 3 weeks.

Post-harvest Management

- Cleaning of the fruits.
- Pre-cooling the fruits.
- Sorting (remove undersized, diseased and damaged fruits).
- Grading: According to shape, size, colour and maturity.
- Packaging: Corrugated fibre board boxes and bamboo baskets.
- Wax coating @ 3 per cent enhances storage life as well as delays ripening.
- Transportation and marketing.

Table 4.3: Insect Pest Management

Insect	*Damage*	*Control*
Mango hopper (*Amirtodus atkinsoni*)	Nymphs and adults suck sap from tender shoots and panicles. Adversely affect the fruit set.	Malathion @ 0.15 per cent.
Mealybug (*Drosicha mangiferae*)	Nymphs suck sap from young shoots, panicles and flower pedicels.	Slippery bands of alkathene around the tree trunk.
Stem borer (*Batocera rufomaculata*)	They are tunnelling through the main trunk or its branches.	Close and plug the entrance of tunnels through kerosene, formalin *etc.*
Shoot gall maker (*Apsylla cistellata*)	Nymphs enter the axillary and terminal buds and turn them into hard conical galls.	Parathion @ 0.04 per cent or Metasystox @ 0.1 per cent.
Stone weevil (*Sternochetus mangiferae*)	Damage pulp and cotyledons of the stone.	General cleanliness.
Fruit fly (*Bactrocera cosyra*)	Lay eggs under the fruit skin. Affected fruits begin to rot and drop down.	Vapour heat treatment, hot water treatment @ 43 to 49 degrees celsius for 8 hours.

Table 4.4: Disease Management

Disease & causal organism	*Damage*	*Management*
Powdery mildew (*Oidium mangiferae*)	Superficial powdery mass/ fungus appears on the leaves, shoots and fruits. Flowers and immature fruits drop in the advanced stage of disease.	Kerathane @ 0.1 per cent.
Anthracnose (*Colletotrichum gloeosporioides*)	Dark brown to black coloured spots appear on the leaves, shoots and fruits. Infected flowers are unable to set fruits and fruits become infected during storage.	Blitox @ 0.3 per cent, Bavistin @ 0.1 per cent.
Sooty mould (*Meliola mangiferae*)	Black sooty moulds develop on honeydew excreted by insects at plant parts and in severe cases plants turn completely black.	Wettable sulphur + Metacid + Gum acacia.
Blight (*Drechslera hawaiiensis*)	Formation of irregularly shaped scattered spots on the leaves.	Spray 0.3 per cent Captan + 0.1 per cent carbendazim + 0.1 per cent thiophanate methyl.

Table 4.5: Physiological Disorders

Disorder	*Characteristics/Cause*	*Control*
Malformation	Compact mass of flowers, sturdy and bear only male flowers.	NAA @ 300 ppm during 1st week of October and pruning of malformed shoots.
Black tip	Black depression at the distal end of fruit .	Borax @ 0.6 per cent.
Soft nose tip	Yellowing of fruit tip and pulp softening in the distal region due to Ca deficiency.	Application of $CaCO_3$ and calcium nitrate.
Internal necrosis	Water-soaked spots on the lower region of fruits.	Borax @ 1 per cent.
Alternate bearing	Bearing in on-off pattern.	Paclobutrazol @ 5-10 g a.i./tree.
Fruit drop	Competition for photosynthates, low levels of auxin, gibberellin, and lack of irrigation.	Mulching and irrigation, NAA @ 50 ppm at pea stage or 2-4, D @ 20 ppm.
Spongy tissue	It was discovered by Cheema and Dhani (1934). Convective heat, high solar radiation and direct exposure of fruits to the sun.	Harvesting at 3/4th matured stage.

Mango Malformation

- ☆ It was first reported from Darbhanga, Bihar, in 1891 by Maries. The malformation is widely prevalent in northern India. The malformed panicles remain unproductive and are characterized by a compact mass of male flowers, greenish in colour and stunted growth. The malformed panicles remain intact with the trees for a considerable period.
- ☆ The complexity of the disorder is attributed to cultural practices, nutritional factors and many other factors like mites, fungal infections, viral infections and hormonal imbalances. The exact cause and control of this malady is yet to be known.
- ☆ However, some remedial measures are recommended as follows:
  - ☆ Pruning of shoots bearing malformed panicles.
  - ☆ Deblossoming of early emerged/infected panicles.
  - ☆ Spray 2 g Captan + 1 ml malathion (July- September).
  - ☆ Spray 300 ppm NAA (September).
  - ☆ Grow resistant cultivar *e.g.* Illaichi, Saisugartha.

Biennial/ Irregular Bearing

- ☆ On year and off year bearing at an alternate year or with more gap.
- ☆ Most of the commercial varieties of North India, *viz.* Dashehari, Langra and Chausa are biennial bearers, while South Indian varieties like Totapuri, Neelum and Bangalora are regular bearers.
- ☆ Climatological factors: High RH (relative humidity), rain and low temperature.
- ☆ Age and size of shoot: 8 – 10 months shoots bear flowers.
- ☆ C:N ratio: Appropriate C:N ratio is 3:2, while high N promotes more vegetative growth.
- ☆ Hormonal balance: Low gibberellic acid and high auxin like substances and inhibitors favour or induce flowering.
- ☆ Proper maintenance of orchards by way of effective control of pests and diseases.
- ☆ Soil application of paclobutrazol (PP333) @ 4 - 5 g per tree in September resulted in early flowering, higher fruit set and the yield. It may be applied every year for regular fruiting, particularly in young trees.
- ☆ Grow regular-bearing cultivars.
- ☆ Deblossom/prune the infected shoots/ branches.
- ☆ Apply potassium nitrate @ 2% and NAA @ 40 ppm during October to induce flowering.

Fruit Drop

- ☆ Type of fruit drops: (i) Pinhead drop, (ii) Post-setting drop and (iii) May-month drop.
- ☆ Embryo abortion, climatic factors, disturbed water relations, lack of nutrition, disease, pest and hormonal imbalances are the major factors that lead to fruit drop.
- ☆ The foliar application of 2,4- D @ 10 - 20 ppm or NAA 20 @ ppm, GA_3 @ 50 ppm is found effective to control fruit drop in mango.
- ☆ Spray 500 g zinc sulphate + 1.5 kg urea in 100 litres of water.

Black Tip

- ☆ It is initiated at the marble stage with a characteristic of yellowing of tissues at the distal end. Gradually, the colour intensifies into black. It was observed by Woodhouse in 1909.
- ☆ Gases like carbon monoxide, sulphur dioxide and ethylene constituting fumes of the brick kilns are known to damage the growing tip of fruits and give rise to symptoms of the black tip.
- ☆ Planting of mango orchards in North-South direction.
- ☆ The orchard should be at least 1.6 km away from a brick kiln of 15 to 18 m height.
- ☆ Spray Borax (0.6 %) at the pea stage, followed by two more sprays at 15-day intervals.
- ☆ Spray copper oxychloride @ 1.5 kg/500 litres of water.

Spongy Tissue

- ☆ It was discovered by Cheema and Dhani in 1934.
- ☆ Alphonso is susceptible variety.
- ☆ Fruits are non-edible. Sour, yellowish and sponge like patches with or without air pockets develop in fruit during ripening.

Management

- ☆ Mulching: Organic or inorganic.
- ☆ Early harvesting of fruits.
- ☆ Grow resistant varieties like Ratna, Arka Aruna and Arka Puneet.

Chapter 5

Banana (*Musa paradisiaca* L.)

Family: Musaceae, 2n = 22, 33 (3x), 44 (4x)

Introduction

- Banana is an important fruit of the tropics. It is a monocotyledon herbaceous plant. True stems remain underground, which is known as rhizome. Botanically, the rhizome is a modified form of the stem.
- The portion above the ground made of sheaths of leaves is known as pseudostem. About 40 leaves are formed till flowering.
- The fruit is seedless and developed without pollination. Auxin stimulation is believed to take part in the development of fruit (*i.e.* parthenocarpy).
- The end of the inflorescence 'pendant' is cooked as a vegetable in Bengal.
- India ranks first in the world in area (15.4 %) under banana production.
- India is the largest producer of banana in the world. Its share is 26.6 per cent of the global banana production.
- Banana is known by Adam's fig, Tree of Wisdom, Tree of Paradise and Apple of Paradise.
- Banana has multiple uses, so it is popularly known as Kalpatharu (a plant with virtue).
- Ripe fruits are delicious and are used for table purposes. The immature fruits are used for vegetable.
- Hybridization work on banana is being conducted at Central Banana Research Station, Adhuthurai, Tamil Nadu.

- In India, major banana-producing states are Andhra Pradesh (16.2 %), Gujarat (14.5 %), Maharashtra, Tamil Nadu, Uttar Pradesh, Karnataka and Madhya Pradesh *etc.*

Origin

- The origin of edible banana is believed to be in the hot, tropical regions of South-East Asia (Assam, Burma-Indo China region).
- The primary centre of origin of table banana (*M. acuminata*) is the Malaysian region, whereas the cooking banana (*M. balbisiana*) originated in Southern India.
- Edible Bananas have arisen as a result of natural crosses between two wild progenitors, *Musa acuminata* and *Musa balbisiana.*

Soil and Climate

- Its cultivation is confined to 30°N to 30°S of the equator.
- Banana can be grown on almost all types of soil, providing an adequate amount of soil moisture. Deep, well-drained, loamy soil with adequate organic matter is ideal for its cultivation.
- The optimum soil pH is 6.5 to 7.5, but it can be grown up to 8.5 with suitable amendments.
- The plant has a restricted root zone, so soil depth and drainage are the two most important factors in the selection of a site for banana cultivation.
- The minimum soil depth for banana cultivation is about a meter.
- Banana crop requires large quantities of water but are susceptible to water stagnation.
- Banana is a tropical plant that requires warm humid climates. It grows well in the regions with temperature ranges from 10 to 40°C, but the optimum temperature is 20 to 30°C for getting a good yield.
- Low temperature, *i.e.* less than 10°C, leads to a condition called choke or impeded inflorescence.
- Frost is the limiting factor for banana cultivation.
- Stagnation of water is harmful and can cause Panama wilt.
- Winds are major constraints in banana production, causing uprooting and collapse of plants. Adjust the time of planting to avoid bunch formation during high wind periods.

Genomics Classification

The family Musaceae has three genera, *viz., Ensete, Musa* and *Musella,* with about 50 species.

1. *Ensete*: It has about 6 to 7 species.
2. *Musa:* It has 40 to 45 species, divided into 5 sections.

i. **Eumusa (2n=22):** Largest section with 13 to 15 species, all are cultivated forms; gives edible parthenocarpic banana and are derived from 2 wild species; *Musa acuminata* (A) and *Musa balbisiana* (B).

ii. **Callimusa (2n=10):** It has 5-6 species; plants are dwarf, suited for ornamental purposes and parthenocarpy is completely absent.

iii. **Australimusa (2n=20):** Fruits are parthenocarpic and bunches are erect, having red sap, the skin turns orange while the fruit ripes.

iv. **Rhodochlamys (2n=22):** It has 5-7 species, spreads from India to Indonesia, pseudostem dwarf (<3 m), erect inflorescence and parthenocarpy absent.

v. **Incertae sedis (x=7; 2n=14):** Tallest in Musaceae family (>10 m).

3. *Musella*: Golden lotus banana, Chinese banana/Chinese yellow banana.

Simmonds and Shepherd (1955) have distinguished the major morphological characters of *M. acuminata* and *M. balbisiana*, which are as follows:

Table 5.1: Morphological Characteristics of *Musa* spp.

Sr.No.	*Characters*	*Musa acuminata*	*Musa balbisiana*
1.	Colour of pseudostem	Heavily marked with black or brown blotches.	Blotches slightly or absent.
2.	Peduncle	Usually downy or hairy.	Glabrous
3.	Pedicel	Short	Long
4.	Ovules	Two regular rows in each loculus.	Four irregular rows in each loculus.
5.	Bract curling	Bracts roll after opening.	Bracts lift but do not roll.
6.	Bract shape	Lanceolate or narrowly ovate, tapering sharply.	Broadly ovate, not tapering sharply.
7.	Bract apex	Acute	Obtuse
8.	Bract colour	Red, dull purple or yellow outside, pink, dull purple inside.	Brownish purple outside, bright crimson inside.
9.	Male flowers colour	Creamy white.	Variably flushed with pink colour.
10.	Stigma colours	Orange or rich yellow.	Cream, pale yellow.

- ☆ Simmonds and Shepherd used a scoring technique to indicate the relative contribution of two wild species to the contribution of cultivars.
- ☆ They identified 15 distinguishing factors between *M. acuminata* and *M. balbisiana*. For each character which agreed with *M. acuminata* scored one and for each character which agreed with *M. balbisiana* scored 5, and intermediate expression of characters were assigned scores 2,3,

or 4 according to their identity. So, score ranges between 15 (15x1) for *M. acuminata* and 75 (15x5) for *M. balbisiana*. An analysis showed that cultivars belonged to six groups (2 diploid, 3 triploid and one tetraploid).

Table 5.2: Scoring of *Musa* Genomes

Group	*Ploidy*	*Score*	*Contribution*	*Nomenclature*	*Example*
1	2x	16-23	AA	Musa AA group	Matti
2	3x	15-21	AAA	Musa AAA group	Gross Michel
	4x	15-20	AAAA*	Musa AAAA group	Bodles Altaforte
3	2x	46-49	AB	Musa AB group	Ney Poovan
4	3x	26-46	AAB	Musa AAB group	Champa
	3x	59-63	ABB	Musa ABB group	Kanchkela
5	4x	63-69	ABBB	Musa ABBB group	Klue Tiparot
6	2x/3x	70-75	BB/BBB	Musa BB/BBB group	Saba

* Not encountered in nature.

- ☆ Bodles Altafort is a synthetic hybrid evaluated in the West Indies.
- ☆ The best-known bananas all over the world for commercial purpose belongs to the pure *acuminata* (AAA) group, but *bulbisiana* genome has drought hardiness and resistance to disease. Therefore, the contributions of hybrids of AB, AAB and ABB are better adapted in monsoons with marked dry seasons.
- ☆ The basic chromosome number of the *Eumusa* section is 11. Edible banana of this genus has 22, 33 and 44 chromosomes, and they are diploid, triploid and tetraploid, respectively.
- ☆ Most of the cultivated types are triploid and have sterile pollens. Diploids are less and tetraploids are rare.

Cultivars

- ☆ **Lady Finger (AB):** It is diploid banana variety. It is a thin-skinned and sweet banana. It is also known as sugar banana, date banana or fig banana. It has a strong root system, so resistant to winds but susceptible to drought.
- ☆ **Dwarf Cavendish (Musa AAA):** Plant stature is dwarf. It is used for table and processing purposes. It is highly susceptible to Sigatoka leaf spot disease. Yield per bunch is 55-60 kg.
- ☆ **Harichhal (Robusta, Musa AAA):** It is semi-tall variety. Fruit are large and delicious, skin is greenish to dull yellow in colour. Keeping quality is good.

- **Lalkela (Musa AAA):** It is grown throughout the world. Plant parts like pseudostem, petiole, mid-rib and fruit rind are purplish red in colour. It is susceptible to bunchy top, nematode and fusarium wilt.
- **Grand Naine or G 9 (Musa AAA):** It is a mutant of Dwarf Cavendish. Fingers are bigger in size with a heavy bunch.
- **Lal Velchi (Musa AAA):** It is also known as Red Banana. The colour of pseudostem, petiole and fruit rind is purple-red. It is highly susceptible to bunchy top, nematode and fusarium wilt. It is grown for red skin.
- **Poovan (Musa AAB group):** It is commercially cultivated for the leaf industry. The fruit has a typical sour, sweet aroma and an attractive golden colour at ripening.
- **Nandran (Musa AAB):** It is a popular variety in Kerala and is used for processing purposes. Fruits remain starchy even at ripening. It is susceptible to banana bract mosaic virus, nematode and borer.
- **Hill Banana (Musa AAB):** It is popular in Tamil Nadu and fruits are better in taste.
- **Rasthali (Musa AAB):** It is a table-purpose cultivar. Fruits have a good aroma and are very tasty. It is susceptible to fusarium wilt.
- **Kanch Kela (Musa ABB):** It is an important cultivar for culinary purposes. It is a very hardy variety and can be grown successfully under rainfed conditions.
- **Monthan (Musa ABB):** It is used for processing purpose and skin usually remains green. It is salt tolerant but susceptible to fusarium wilt. It is used for culinary purposes.
- **CO. 1 Banana:** It is a hybrid from Tamil Nadu and the bunch's weight is about 10 kg.
- Giant Governor (*Musa* AAA) (It belongs to the Cavendish family), H1, H2, FHIA 1 (Goldfinger), Kaveri Saba, Kaveri Kalki, Kaveri Sugantham, Kaveri Kanya, Phule Pride and Kaveri Haritha.

Propagation

- Banana propagated vegetatively using suckers (sword suckers/water suckers), rhizome and tissue culture plants.
- Sword suckers are better for propagation (weight about 750 g).
- In addition to the above, 'Bits' and 'Peepers' (a cut portion of rhizomes) are used for propagation.
- It is parthenocarpic in nature, so sexual propagation is not possible.

Planting

- The time of planting depends upon cultivars, agro-climatic conditions and market demand. Banana can be planted throughout the year, except in severe winter and heavy rains. However, most commonly, it is planted during June-July.

Planting Time in Different areas is as under:

- The area of heavy and continuous rains is September-October.
- The area of limited rains is June-July.
- The area of assured irrigation is February-March.
- The area of hill slopes of South India is February-March.
- Banks of Cauvery River is April.

Planting Method

- Pit (60 cm^3) or trench methods are commonly followed.
- Spacing for tall cultivars is 2.7 m x 3.0 m.
- Spacing for medium sized cultivars is 2.1 m x 2.1 m.
- Spacing for dwarf cultivars is 1.8 m x 1.8 m.
- For high-density planting, Robusta and Dwarf Cavendish spaced at 1.5 m x 1.5 m accommodate 4444 plants/ha (recommended by IIHR).

Manures and Fertilizers

- Banana is a heavy feeder crop due to its rapid and vigorous growth as well as high fruit yield, so it requires a large quantity of nutrients.
- The application of nutrients depends upon soil type, cultivars, agroclimatic, conditions and production systems.
- Apply nitrogen @ 100-250 g per plant in 3-4 split doses, *i.e.* 2nd, 4th and 6th months after planting (add 25 per cent through FYM).
- Phosphorus requirement at the time of planting is low, 40-60 g per plant.
- Potash application is very essential, 200-300 g K_2O per plant in two split doses at vegetative and reproductive phase.
- The fertilizers should be applied in a basin of a 30-45 cm radius from the plant and mix it with the soil and irrigate just after fertilizer application. Fulfill the fertilizer requirement before flower initiation, which normally occurs 5-6 months after planting.

Irrigation

- Banana is a moisture-loving plant. The water requirement of banana is very high, therefore, if there is no rain, the plant should be irrigated immediately after planting and the soil in banana plantations should not be allowed to dry completely throughout the year.
- Drip irrigation can reduce the 40-50 per cent of water requirement and increase yield and leaf production, which decreases the number of days to harvest.
- About 40-50 irrigations (1800-2200 mm) are required from planting to harvest in island areas. ABB clones can be grown in rainfed areas.

Mulching: Organic and inorganic mulching reduces water requirements. It control weeds and maintains the soil temperature.

Furrow method: This method also reduced water wastage.

Important Cultural Practices

(i) **Desuckering:** As banana produces a number of suckers if allowed, they compete for moisture and nutrition with the mother plant; one or two healthy sword suckers may be allowed to grow for a ratoon crop. Desuckering once in 45 days is very common.

(ii) **Mattocking:** The plant pseudostem is cut 30-45 days after harvest to facilitate nutrient mobilization from the mother plant to the developing ratoon plant. Pseudostem should be cut, leaving a stump of 0.6 m in height.

(iii) **Earthing up:** Soil is mounded around the pseudostem during the rainy season to protect the plants from water logging. It also provides support to the base of the plant to prevent uprooting and help in the formation of good root system.

(iv) **Propping:** Provides support to the plant at the bunch stage with bamboo crutches. If the variety is dwarf and planted at closed spacing, no propping is required.

(v) **Mulching:** Banana crops respond to mulching. It reduces the cost of cultivation by reducing the number of irrigations and weed growth.

(vi) **Denaveling:** Removal of male buds after setting of last fruit. It serves a dual purpose *viz.*, avoid the movement of food into an unwanted sink and these buds can be sold as vegetables.

(vii) **Frost protection:** In North India frost from December to January is a severe problem. Smoking and the application of light irrigation decrease the effect of frost. But smoking should be avoided to save the environment.

(viii) **Thrashing:** A process of removal of old, dry, diseased and senescent leaves. Minimum 12 leaves are to be retained for maximum yields.

(ix) **Bunch covering/wrapping:** Bagging of the bunch with perforated polythene cover or dried leaves to protect against cold, sun scorching, attack of thrips and other insects during the bunch maturity stage. Bags may be coated with pesticides.

(x) **Ratooning:** Banana yields three crops from a single time planting, which means one main crop and two ratoon crops.

Intercrop

- Intercropping: Intercrops like turmeric, ginger, radish, cabbage, spinach, soya bean, cowpea and onion can be taken.
- Mixed cropping of banana with coconut, areca nut and cassava is common in coastal areas of Tamil Nadu.

Harvesting

- It is a climacteric fruit, so it is harvested at a green, fully mature stage when the top leaves start drying.
- The colour changes from deep green to light green.
- Shedding of floral end of the fruits by a soft touch with the hand is an indication of maturity.
- Harvest by cutting the bunch with a 15-25 cm stalk for handling.
- Under favourable conditions, banana start flowering in 9-12 months and fruits mature in about 4-5 months, depending upon varieties and climate.
- Fruits are harvested at the $\frac{3}{4}^{th}$ maturity stage for distant markets or for chip-making purposes while, for local markets, fruits are harvested at full maturity.

Ripening

- Smoke treatment by straw, leaves or cow dung is common in Tamil Nadu and Maharashtra in a closed chamber, fruits ripen in 18-24 hours during summer and 48 hours during winter, and after that, shift fruits in a ventilated room for uniform ripening.
- Dip treatment in 500 ppm Ethephon for 2-3 minutes, then air dried and kept at 16-18°C with 90-95 per cent relative humidity (RH) for uniform colour development and excellent ripening.
- Apply ethylene @ 100 ppm in a closed chamber for 24 hrs for uniform colour development and ripening.

Storage

- Fruits can be stored at 13 °C and 80-85 per cent relative humidity (RH) for three weeks.
- At low temperatures, the fruits become black. Therefore, these should not be kept in the refrigerator.
- Fruit ripening can be delayed, or shelf life can be increased by GA_3 @ 150 ppm + dip in 6 per cent waxol.
- Use ethylene absorbent (potassium permanganate) to enhance the shelf life of fruits.
- The storage life can be increased by keeping the fruits in high concentrations of carbon dioxide and low concentrations of oxygen.

Insects

1. **Pseudostem borer (Pseudostem weevil)**: Adults of female weevil lays egg on the pseudostem. Grub emerge and feeds the pseudostem extensively. During severe attacks leaves turn yellow and the plant dies.

   Remove dried leaves, use healthy planting material and prune the side suckers every month.

2. **Rhizome weevil**: The grubs of weevil make tunnels and feed the rhizome and leaves, which reduce fruit yield. Malbhog, Champa and Nendran are highly susceptible.

   Dip suckers in 0.1 per cent quinalphos solution.

3. **Banana aphid**: It acts as a vector for transmission of the bunchy top virus of banana.

   Apply 0.3 per cent Rogor.

4. **Fruit and leaf scarring beetle/Flea beetle/Banana beetle**: Beetle larvae roll the central leaf. It feeds on leaves, flowers and peels of newly emerged fruits. It forms irregular patches on furled and unfurled leaves.

   Use night trape, pheromone trape or dip the beetles in Kerosene oil.

Diseases

1. **Panama wilt:** It is a serious soil-borne fungus reported from Panama caused by *Fusarium oxysporium F.* sp *cubense.* Its damage results in the yellowing of leaves, rotting of pseudostem and discolouration of vascular strands. Poorly drained acidic soils with high nitrogen application are favouring the disease.

   Rasthali is highly susceptible, whereas Robusta and Dwarf Cavendish are resistant varieties. Apply Bavistin 0.1 per cent during the rainy season.

2. **Leaf spot/Sigatoka:** This disease was first reported from the Sigatoka Valley, Fiji (1913). Initially, yellowing streaks appear on the leaves, later on, these enlarge into cylindrical shapes. The central portion of the spots becomes grey with a brownish margin and the leaf dries off in the advanced stage.

   The Gross Michel and Cavendish group are (AAA) highly susceptible to Sigatoka. While all ABB clones are resistant to leaf spots. Spray 0.2 per cent Dithane M-45.

3. **Banana bunchy top virus (BBTV):** It is also called cabbage top. This disease was first observed in Fiji in 1891. It is a viral disease transmitted by banana aphids. The initial symptoms are the appearance of a dark green streak on the petioles and veins. The internodal space and leaf size decreased. Later on, it looks like a bunch of crowded leaves at the apex and the fruit formation is very small or like a figure. Dwarf banana cultivars are very susceptible.

   Control the population of aphids.

4. **Leaf blight:** Formation of a yellow spot on the leaves. In the advanced stage, these spots enlarge and turn brown.

   Spray 0.1 per cent Bavistin.

5. **Stem end rot:** Water-soaked spots appear on the stem/ rhizome and tissues become soft. Pathogen spreads over the fruit and causes black discolouration of fruit skin.

   Spray 0.1 per cent Bavistin.

6. **Cigar end rot:** It looks like a ash on the end of the cigar. Figures on some branches or immature fruits are infected (blackening of skin), however, in some branches, all figures are infected.

   Spray copper oxychloride @ 0.25 per cent. Cover the flowers before the emergence of the head.

Physiological Disorder

Neer vazhai: Fruit oozes out watery fluid when cut, hence the name 'neer' means water and 'vazhai' means banana. Infected plants show poor growth, delayed shooting (flowering) and immature unfilled fingers, Narendran banana is most affected in Tamil Nadu. Causes are not known but transmitted through suckers. It may be due to viral infestation. NAA improves the finger filling.

Kotta vazhai: Causes are not known. 'Kotta' means seed 'vazhai' means banana. It refers to enlarged ovules and immature dark green fruits. Application of 2,4-D @ 120 ppm enables the plant to obtain a normal bunch. It is a very serious disorder in the Poovan variety.

Chapter 6

Ber (*Ziziphus mauritiana* Lamk.): King of Arid Fruits

Family: Rhamnaceae; 2n = 24, 48 (4x)

Introduction

- Ber/Indian jujube is a hardy, one of India and China's ancient fruits. It is king of arid fruits, also called poor man's fruit/poor man's apple. However, grafted varieties of ber are no longer the poor man's fruit as they fetch even higher prices than several other fruits.
- Ber/Indian jujube is one of the ancient fruit crops popular due to its nutritional value and good sugar: acid blending in taste.
- India ranks 2nd after China in area and production of ber.
- Its fruit is a rich source of vitamins and minerals. It contains vitamin C, about 90-120 mg/100 g pulp, and is a good source of vitamins A and B. Vitamin C content in the pulp is higher near the stone as compared to vitamin C content in the pulp near the skin.
- Fully ripe fruits of ber contain 13-19°B TSS, 0.2 to 0.6 per cent acidity and 0.94 per cent proteins. These constituents vary with the cultivars and growing conditions.
- A comparison of the nutritional value of ber with apple reveals that it excels in calcium, phosphorus, protein, mineral matter, vitamin C and vitamin A. All the parts of the plant, such as the seed, stem, bark and roots contain alkaloids.

- Mostly ber fruits are consumed as fresh, and these can also be used for different value-added products such as dried ber/chhuhara, preserve, pickle and other beverages like juice, ready-to-serve (RTS), squash and syrup.
- The ber extract helps in digestion and is said to be a blood purifier. Powder from root bark is effective to cure ulcers, fever and wounds.
- Its fruits are considered safe and nutritious during pregnancy and to decrease cholesterol levels and blood pressure. Fruits are also advised as a fighting agent against HIV and cancer. It is also prescribed during open heart surgery for fast healing and recovery.
- The bark of the stem is recommended or prescribed during diarrhoea. Its leaves are used as fodder/feed for animals; branches are used for fencing and wood is used as timber and to make parts of agricultural implements. Ber tree can be used to rare the lac insects.
- Ber is the main crop in India, but it is also grown in China, Taiwan and the central part of Asia. In India, it is mainly grown in Rajasthan, Haryana, Punjab, Gujarat, Madhya Pradesh, Maharashtra *etc.* In addition to the above states, it is grown in pockets of arid zones of India.

Origin

Indigenous to the area stretching from India to South Western China and Malaysia.

Soil and Climate

This can be grown on all kinds of soils in which other crops cannot be grown easily because it is a hardy crop with the tap root system; however, good crops can be achieved in fertile, deep and well-drained soils. Slightly saline to alkaline soils are suitable for its growth. The plant can survive easily under high pH soils (9.2). This crop performs well in slightly saline to alkaline soils.

Ber is distributed between 30°S and 30°N latitude. Ber is exceptionally hardy and grows successfully even under the conditions of unfavourable climates where most of the other fruits fail to grow. It prefers a hot and dry climate (arid to semi-arid) for quality fruits and production. It survives well under extremely hot conditions but not in frost conditions. It sheds leaves and enters into dormancy during summer (May-June). New growth flesh starts with the onset of monsoon, *i.e.* the last week of June or 1st week of July, and it continues up to flowering. The leaves of ber have thick cuticles and sunken stomata, which results in less transpiration from the plant system. Buds are covered with leaf scales; therefore, it adapt to dry climates. It has xerophytic characteristics, so it survives well under extreme drought conditions. High humidity is harmful to fruit development.

Species and Cultivars

There are different species grown in different parts of the World.

Indian

- *Z. mauritiana,* fruits are edible, leaves are used for fodder purposes and its bark is used for tanning purposes.
- *Z. rotundifolia:* Found in the south western part of India, fruits are edible, leaves are used as fodder and bark is used for tanning purposes.
- *Z. nummularia:* It is a herbaceous plant grown in the south western part of India. Fruits are edible, leaves are used as fodder and bark is used for tanning purposes.

Chinese: *Ziziphus jujuba,* fruits are edible, its leaves are used for fodder and bark is used for tanning purposes.

Iran and Iraq: *Z. spina.*

Cultivars

On the basis of maturity time ber cultivars are classified into three groups:

1. **Early maturing (Mature during February)**: Gola, Seb, Sandhura Narnaul.
2. **Mid maturing (Mature during 3rd week of February to 3rd week of March):** Kaithali, Mudia Murhara, Banarsi Karaka, Sanaur -5, Chhuhara.
3. **Late maturing (Mature during 2nd week of March to April)**: Umran, Illaichi, Kathaphal, Ponda.

There are more than 125 cultivars developed by selection and grown in different regions of India.

Early Cultivars

1. **Gola:** Round-shaped shiny fruit, the weight of fruit is 14.6 g, TSS is 17.4°B, acidity is 0.46 per cent, vitamin-C is 85.5 mg/100 g pulp. The average yield per tree is 100 kg. The fruits ripen in the first week of February.
2. **Seb:** A good early ripening cultivar. Fruit weight is 14 g, TSS is 20.7 °B and acidity is 0.44 per cent. Average yield per tree is 80 kg, vitamin C is 85.0 mg/100 g pulp.
3. **Sandhura Narnaul:** The size of the fruit is long and pointed, colour- greenish yellow, fruit weight is 15.7 g, TSS is 16.8°B, acidity is 0.98 per cent, vitamin C is 87.5 mg/100 g pulp, average yield per tree is 85 kg.

Mid-Season Cultivars

1. **Banarsi Karaka:** A mid-season cultivar. Fruits are long and pointed, and the fruit colour at ripening is yellowish. TSS is 16.9°B, acidity is 0.13 per cent, vitamin C is 110 mg/100 g pulp, average yield per tree is 125 kg.
2. **Kaithali:** The average fruit weight is 17.8 g, TSS 17.6°B, acidity 0.51 per cent, vitamin C 98.3 mg/ 100 g pulp. The yield is 125 kg per tree. This is a very good variety for rainfed areas.

3. **Muria Murhara:** The fruits are bell shape and become yellowish at the time of ripening. The fruit weight is 24 g, TSS is 18.5°B, acidity is 0.29 per cent and vitamin C is 90.7 mg/100 g pulp. The average yield per tree is 125 kg.
4. **Sanauri No. 5:** Fruits of this variety are oval in shape with pointed heads and fruit weight is 17.6 g, TSS of fruit is 18.1°B, acidity is 0.26 per cent, vitamin C is 70 mg/100 g pulp pulp and fruit yield is 100 kg/plant.
5. **Chhuhara:** Fruits are long and deep green. The skin of the fruit is rough. The average weight of the fruit is 22.25 g, TSS is 18.5°B, acidity is 0.26 per cent, vitamin C is 70 mg/100 g pulp. The average yield per tree is about 100 kg.

Late Season Cultivars

1. **Umran:** Fruit is a large-sized and high-yielding cultivar. The average weight per fruit is 35 g, skin of the fruit is thick and hard, TSS is 19.5°B, acidity is 0.33 per cent, vitamin C is 80 mg/100 g pulp and the yield per tree is 200 kg.
2. **Katha Phal:** Fruits are purple coloured with pulp granular like apples. The raw fruit is light red green in colour and at ripening, it becomes purple to brown in colour. The average fruit weight is 18 g, TSS is 16 °B, acidity is 0.76 per cent. Vitamin C is 84.6 mg/100 g pulp and fruit yield is 120 kg/tree.

3 **Illaichi:** This is a promising variety with very small-sized stone. Its fruits are available throughout March. Fruit is round and small and its colour becomes brown at ripening. The fruit weight is 6.48 g. TSS is 21°B acidity is 0.23 per cent, vitamin C is 124.6 mg/100 g pulp and fruit yield is 100 kg/tree. This variety is pollen sterile variety (90 %).

Other Important Varieties and Hybrids

1. **Thar Sevika or CIAH Hybrid -1 (Seb x Katha):** It is an early maturing cultivar. The average fruit weight is 25 g, TSS is 24.3 °B, ascorbic acid is 88 mg/100 g pulp. Fruit yield from five year old plant's is 28-30 kg/tree.
2. **Thar Bhubraj (CIAH Selection-1):** It is an early maturing variety. The average fruit weight is 27 g, TSS is 23 °B and ascorbic acid content is 60 mg/100 g of pulp. Five years old plant produces about 35-35 kg of fruits.
3. **Goma Kriti (CHES-I):** Clonal selection from Umran in Gujarat. It is early maturing, about 3 weeks earlier than Umran. Fruit-keeping quality is suitable for distant markets. It is less affected by fruit flies, fruit borer and powdery mildew.
4. **Narendra Ber Selection 1:** Released from ANDUA & T, Ayodhya, UP.
5. **Narendra Ber Selection 2:** Released from ANDUA & T, Ayodhya, UP.
6. **Thar Malti:** Released from CIAH Bikaner, Rajasthan.

7. **CAZRI Ber 2018:** Early maturing and high yielding. Its yield is 25 per cent higher than Gola under rainfed conditions.
8. **Phule Shabari (SBL 26):** Tolerant to fruit borer and powdery mildew disease.

Propagation

Most common method for propagation is a shield (T-Budding) budding in July and August. Other methods are ring budding and I-budding. Budding should be done 15-20 cm above the ground level. The time taken for successful sprouting of bud depends on temperature, humidity and quality of bud used. In North India, July-September is the most suitable period for successful budding. Under *In situ* conditions, the budding success rate is more due to well established roots system.

Rootstocks

Z. rotundifolia and *Z. nummularia* (slow growing), Kakrol rootstock are used at the Experimental orchard of CCSHAU, RRS, Bawal, Rewari. Other rootstocks are Tikari and Sukhavani.

Raising rootstocks: Crack hard seed coat (scarification) and sowing in April. Seedlings can be raised in polythene bags (30 cm x 10 cm size) in the month of April. Improvement in seed germination is suggested as under:

1. Soak the stones for 48 hrs in normal water or 15-20 minutes in hot water.
2. Soak the stones in concentrated H_2SO_4 for 6 minutes, followed by washing with an alkaline solution.
3. Dip the stone in GA_3 @ 200 ppm for 6 hours.

Ber seeds have hard covering, so it is difficult to grow and take about 3-4 weeks to emerge seedlings. Carefully cracking the endocarp reduced the germination period by one week. The seedling plants are used as rootstocks.

Planting

A planting at 8 m to 10 m spacing is followed for ber, however, under rainfed and dry conditions, ber plantation can be made at 6 m x 6 m distance by using semi-vigorous rootstock (*Z. mauritiana*, Coimbatore). The monsoon period (August-September) and February-March are the best time for transplanting. Bare-rooted transplanting can be done during December-January. Dig a pit of 1 m x 1 m x 1 m dimension during May-June or 1-2 months before planting. Add 60 kg FYM+ 2 kg SSP +15 ml chlorpyriphos along with the top 50 per cent soil in each pit. The plant should be watered immediately after plantation. In sandy soils, add bentonite clay to reduce the infiltration of water and increase the success rate. The tree starts giving commercial crops with higher yield and excellent quality fruit from the third year of planting.

After Care of Young Plants

Apply light and frequent irrigation to newly planted orchards. Regularly remove the sprouts emerging from the stock for better growth of scion. Plants should be properly staked for upright growth and protect from strong winds, hot sun and water stagnation. Trees for windbreak can also be planted around the commercial orchard.

Manures and Fertilizers

A deficiency of nutrients, especially N and P, hinders the growth of plants for the production of fruiting branches. It reduces flowering and plants fail to set fruits. For good yield, the application of manures and fertilizers is necessary. On the basis of leaf nutrient status of the 5th and 6th leaf of secondary shoots, analysed to decide nutrient requirements.

Table 6.1: Application of Manures and Fertilizers

Plant age (year)	*Farmyard manure (kg)*	*Urea (kg)*	*Single super phosphate (kg)*
1	10	0.200	0.300
2	10	0.400	0.600
3	20	0.600	0.900
4	25	0.800	1.200
5	30	1.000	1.500
6	40	1.200	2.000
7 and >	50	1.250	2.500

Apply FYM +SSP+ ½ urea in the last week of June or the first week of July after pruning and the remaining ½ urea in November.

Foliar Application

Spray 1.5 per cent urea along with 0.5 per cent $ZnSO_4$ in the months of July and November to improve growth and reduce fruit drop.

Irrigation

Young plants should be irrigated regularly. Well-grown plants have deep root systems, shed leaves during summer, so very little or no irrigation is required. The growth phase coincides with rainfall and fruiting with winter. Four irrigations are useful for better yield and quality, 1st in June, 2nd in November, 3rd and 4th in January. Avoid irrigation from September to mid-November due to flowering. In rainfed areas, water harvesting can be done by giving a 5 per cent slope towards the trunk and mulching to conserve moisture. Ber crop is very hardy and have deep tap roots, and is xerophytic in nature, so once established, it can flourish well for a longer period with little or less care. It requires irrigation from November to February or for early maturing varieties up to January for better fruit development.

Training

When the plant is not pruned and allowed to grow, it will not bear or bear very little fruits of poor quality. Unpruned plants become vigorous and becomes hurdle for intercultural operations. The ber fruit starts bearing in 2-3 years. Hence, it should be properly trained to build a strong framework. In the initial phase, plant is allowed to grow a single stem up to 75-90 cm height from ground level to develop a strong framework. Now-a-days, high-density planting or modern training systems, training to single stem height can be opted as per requirement of the training system. At the initial stage, staking the plant with bamboo sticks is needed. Later, three to five widely spread branches in all directions may be encouraged. The process should be followed up to three years to allow equal/balanced distribution of primary, secondary and tertiary branches on the plant for more yield and better quality production. It will also make a strong framework for the plant. Train the plant according to the modified leader system.

Pruning

Ber requires regular pruning because it bears flowers and fruits on the current season's growth. Regular pruning induces more number of fruiting branches on the plant. So, for more yield and better-quality fruits, regular pruning is required. Remove undesirable, weak, criss-cross, diseased, dried and broken branches. After the establishment of frame work, 1/3rd to 1/4th of the previous season's growth should be pruned (keeping 6-8 buds of the previous season) during the dormancy period, *i.e.* from 15th May to 15th June in North Indian condition. The best time for pruning is the hot, dry season when plants shed their leaves and go to rest or become dormant. Heavy pruning induces excessive vegetative growth, which causes less and delayed flowering as well as reduces fruit yield.

Rejuvenation Old Trees

Primary branches of old, weak and unproductive trees are headed back at 1.5 to 2.0 m during the dormant period (May-June). Apply FYM and irrigation after pruning. The tree should be rejuvenated after the age of 25 years.

Head back to the old ber plant (above 30 years) at 1.5 m above the ground level and keep the six strongest shoots during the next season. These shoots should be further headed back at 90 cm for better yield and growth.

Top Working

Conversion of inferior/well-grown desi ber trees into good-bearing commercial varieties. The sprouts on the headed back branches are budded with desirable scion in July –August.

Flowering

Duration of flowering varies with area, climate and cultivars. Flowering occurs from July to November in different parts of the country. The time of anthesis varies from cultivar to cultivar. In most of the cultivars, anthesis time is during morning, afternoon and evening hours. In North India, flowering takes place from September to November. Ber trees produce two types of flowers- male and hermaphrodite. The percentage of hermaphrodite flowers is low in the early season of flowering but increases with advancement of season. Ber is a cross-pollinated crop and the most common pollinating agents are houseflies, honey bees and yellow wasps. Pollen grains are highly viable. The viability ranges between 87-91 per cent in different cultivars. Fruit set varies from 6.19 to 8.16 per cent. However, under controlled cross-pollination, it ranges from 26 to 46 per cent. Spraying of GA_3; 2,4,5-T; 2,4-D, NAA @ 10-15 ppm are found to improve the fruit set. The application of GA_3 and auxin results in pollen germination and pollen tube growth. Ber has a double sigmoid pattern of fruit as well as stone growth. It is completed in three different phases: 1st phase occurs up to 45 days after the fruit set, 2nd phase occurs from 45 to 90 days after the fruit set and 3rd or final phase occurs from 90 to 185 days after the fruit set. The growth patterns of fruits and seeds are similar. Fruit of ber is a drupe.

Fruit Drop

The major problem of fruit drop is in Illaichi.

Physiological Fruit Drop

1. **Abortion of embryo:** During the early stage, the drop is mainly due to shrivelling of fruits, it occurred up to December. It accounts for about 30 per cent of fruit drop.
2. **Premature fruit drop:** Due to the formation of an abscission layer in the stalk of ripe and nearly ripe fruits. This abscission has been reported as a result of natural exhaustion of growth regulators present in the fruit or when they cease to flow from their production centre (source) because of some disorders in the tree or climatic conditions. Maximum fruit drop occurs from November to end of January (about 60 %). Fruit drop from the first week of February to March remains 17.7 per cent. The synthesis of auxin and gibberellins decreased during the growth period due to the formation of inhibitors. Foliar application of 20-30 ppm NAA once in the second fortnight of October and again in the second fortnight of November check 10 per cent fruit drop in Umran.
3. **Fruit drop due to climatic factors:** Severe cold and poor photosynthesis in December fail to provide the required amount of food substances.
4. **Fruit drop due to white ant attack:** Block the smooth flow of sap in the xylem and phloem. Apply chlorpyriphos 1 ml per litre of water.

Pathological Fruit Drop

Powdery mildew: Apply Karathane @ 0.05 per cent or Sulphex @ 0.2 per cent.

Harvesting

Ber matures in 150-175 days after flowering. A preharvest spray of 750 ppm Ethephon at the colour turning stage induces early maturity. Its fruit is climacteric raspatory behaviour. Harvesting time in different regions is as under:

1. South India: October-November.
2. Gujrat: December-March.
3. Rajasthan: January-March.
4. North India: February-April.

Ber plants of 10 to 20 years old attain full maturity and achieve maximum level of yield, *i.e.* 80 to 200 kg fruits. However, it varies with cultivar and region.

Grading and Packaging

The fruit should be harvested at the colour break stage, *i.e.* greenish to golden yellow colour along with some brownish colour tings. Under-ripe fruits never attain the desired quality and taste, at the same time, over ripe fruits also lose their quality. Under-ripe and over-ripe fruits have the least demand in the market. For better shelf life and high market rates, fruit should be harvested manually. Fruit should be graded on the basis of size.

Storage

- ☆ Ber fruit can be stored at room temperature for 10-12 days in perforated polybags.
- ☆ Dip the ber fruits in 6 per cent waxol solution to decrease physiological loss in weight, decay loss and delayed ripening or increase the shelf life of the fruits.
- ☆ At 0 to 4°C, Umran fruits can be stored for about 3 weeks.
- ☆ A pre-harvest spray of 0.05 per cent Captan or Dithane M-45 increase shelf-life up to 8 days.
- ☆ Cold storage of ber fruits at 3°C with 85 per cent relative humidity extends shelf life up to 30 days.

Insects

1. **Fruit fly:** The adult fly lays eggs on the fruit and larvae feed on fruit pulp. The infested fruit becomes deformed and not fit for consumption.

   Spray 0.1 per cent dimethoate at pea size stage or spray 0.1 per cent malathion + *gur* from the last week of January onward. BS 1 and BS 2 germplasm of ber are resistant to fruit flies.

2. **Lac insect:** It attacks on the shoots and causes deposition of brownish mass on the attached/ infested shoot, later on, these shoots dry off.

   Spray 0.1 per cent quinalphos after the first rainfall.

3. **Leaf-eating caterpillar:** Caterpillar feeds and rolls on leaves and makes hole in the leaves.

   Spray 0.1 per cent quinalphos after the first rainfall.

4. **Termite:** Damage is more severe in newly planted orchards and nurseries. Formation of soil tunnels on plant and serrated cut seen on roots. It feeds on the root and lower stem portion.

   Apply 50 ml chlorpyriphos/plant.

5. **Bark-eating caterpillar:** This insect feeds inside the hole near the branches during the day and it feed outer bark or surrounding of the hole during night.

   Kill the caterpillar with the help of wire or inject insecticide or Kerosene oil in the hole. After that, cover it with mud or cotton.

Diseases

1. **Powdery mildew:** The powdery mildew mass appears on the leaf and fruit surface during November (or up to January under North Indian conditions). Later on, the affected portion of the fruit becomes sunken and corky.

   Spray 0.2 per cent Sulphex or 0.1 per cent Karathane before flowering and after fruit set. Sanur 2, BS 1 and BS 2 germplasm are resistant to powdery mildew.

2. **Sooty mould or Black leaf spot:** The black spot appears on the lower surface of the leaf. Later on, these spots become brownish and leaves fall off.

   Spray 0.3 per cent copper oxychloride.

3. **Fruit rot:** Dark brown spots appear on the fruit surface and fruit size remains small.

   Spray 0.3 per cent copper oxychloride.

Physiological Disorder

Fruit Crack

This can be caused by a rise in temperature in March, irregular irrigation intervals, and an imbalance of endogenous levels of growth regulators and varietal characters. High levels of gibberellins and abscisic acid in skin and stone lead to fruit cracking. Umran has less problem, whereas Gola, Kaithali, Sanaur-2 and ZG-2 are prone to splitting due to soft skin.

Control: Regular irrigation helps in the uptake of water and solute. Spray NAA @ 20-30 ppm at the pit hardening stage once in October and again in the second week of November, which increases the promoter level and thus reduces the skin cracking of fruit.

Chapter 7

Aonla or Indian Gooseberry or Amla (*Emblica officinalis* Gaertn./ *Phyllanthus emblica* L.)

Family: Euphorbiaceae/Phyllanthaceae; 2n = 28

Introduction

- Aonla has been regarded as a sacred tree and known as *Amritphal* in olden literature. It is an important crop of Indian origin grown in arid and semi-arid parts of the country, especially in North-West part of India. Nutritionally, it is very rich in vitamin C (500-1500 mg/100 g pulp), being next to Barbados cherry. Vitamin C in aonla is not oxidized due to presence of the tannins in fruits.
- The plant is quite hardy, heavy bearer and gives better income to farmers.
- *Trifla* and *Chyawanprash* are the most common commercial products of aonla.
- India ranks first in area and production in the world. It is also grown in Sri Lanka, Cuba, USA, Iran, Iraq, Pakistan, China, Malaysia, Bhutan, Thailand, Vietnam *etc.*
- In India, major aonla producing states are Uttar Pradesh (36 %), Madhya Pradesh (28 %), Tamil Nadu, Gujrat, Chhattisgarh, Haryana, Odisha, Jammu & Kashmir and Rajasthan *etc.*

Origin

Aonla is native to tropical south eastern Asia, mainly peninsular of central and southern India.

Soil and Climate

- Sandy loam soils are the best suited for aonla cultivation.
- Soil should be at least 2 m deep. Otherwise, the plant will not prolong beyond 12 years in fruiting.
- Aonla can be cultivated in marginal soils – slightly acidic to saline/sodic (pH up to 9.5, ECe 6.0 to 9.0 dSm^{-1}, ESP 35) conditions.
- Heavy soils or high water table areas are not suited for its cultivation.
- Its cultivation is more successful in subtropical regions. It can also be grown in semi-arid to arid regions.
- Heavy frost during winter is not suitable for its cultivation. A mature aonla tree can tolerate freezing as well as high temperature up to 46°C.
- Dry spells result in heavy dropping and delay in the initiation of fruit growth.
- Stagnation of water during rains is harmful to its growth and root system.

Cultivars

- **Banarasi:** It is an early and heavy bearer cultivar with good quality fruits. It was a seedling selection from Varanasi earlier it was Banaras, so it is named Banarsi.
- **Chakaiya:** It is a hardy and mid-season cultivar with heavy bearing habit. It is a good pollinizer also. Its shape is like 'chakla', so it is named Chakaiya.
- **Francis (Hathijhool):** It is a late maturing and heavy bearer cultivar. Branches are drooping, fruits are oval-shaped and have reddish skin colour.
- **NA 4 (Kanchan):** Seedling selection from Chakaiya. It is a mid-season maturing variety. Fruits are medium-sized and have high fibre content.
- **NA 5 (Krishna):** Selected from Banarsi. It is early maturing. Fruits are triangular and conical shaped with less fibre content.
- **NA 6 (Amrit):** This is a selection from a Chakaiya. It is a mid-season maturing variety. Medium-sized shinning fruits with low fibre content. Fruits are suitable for the processing industry.
- **NA 7 (Neelum):** This is a selection from Francis. It is a mid-season maturing cultivar. It is a prolific bearer and fruit size is medium to large.
- **NA 10 (Balwant):** It is a chance seedling selection from Banarsi. Fruits are medium sized, flattened, round in shape, skin is rough and has high fibre content.

- **NA 20:** Fruits are large-sized and have high fibre content. The average fruit weight under Bawal conditions is 55 g.

Other varieties: Anand 1, Anand 2, Laxmi 52.

Propagation

- Aonla can be propagated both by seed and vegetative propagation.
- Seedling plants are not true to type, bear small-sized fruits and having lower yields per plant. Most of the seedling plants are used as rootstocks.
- For high yield and better quality fruits, plants are propagated vegetatively with elite germplasm/improved cultivar.
- The ripened/matured fruits of aonla are dried in the shade to extract seeds. The fruits and capsules start splitting during the drying process.
- The dried seeds are dipped in salt solution, the seeds settled in the solution are used for sowing and floating seeds are discarded.
- Dip seeds in GA_3 @ 500 ppm for 24 hrs for maximum seed germination.
- Seeds are sown in the nursery during March or June-July.
- The seedling plants are budded during July-August after attaining proper stem girth of pencil thickness.
- Patch budding is followed for vegetative propagation in North India.

Planting

- Planting should be done in July-August and during the spring season (15 January to 15 February).
- Pits of 1 m^3 size are dug 2 months prior to planting. In each pit, fill 50 kg FYM + 2 kg SSP along with ½ top soils.
- In sodic soil, gypsum is added according to requirement.
- Planting space is decided on the basis of the soil, climate and growth habits of the plant. In general, a distance of 7-10 m is followed for planting.
- It is a cross-pollinated crop. Therefore, plants of two or more varieties should be planted together to get a better yield.

Manures and Fertilizers

- Nutrition plays an important role in the cultivation of aonla. Apply 15 kg farmyard manure to one year old plant. This dose is increased almost in a similar way and stabilized at 5-6 years of age.
- Add 500 g urea, 2.5 kg SSP per tree in February and 500 g urea in the month of July to the fully mature tree.
- Narcosis can be controlled by the application of Borax solution @ 0.6 per cent thrice at fortnightly intervals during September– October.

Irrigation

- Aonla is generally cultivated as a rainfed crop. But for better establishment of orchards, young plants require frequent irrigations.
- During the active growth period from March to June, trees are required to be irrigated at weekly intervals for higher fruit sets and reduced fruit drops.
- During the fruit development period from September to October, monthly irrigation should be scheduled.
- During the peak winter, light irrigation should be given to avoid frost injury.
- Avoid irrigation during the flowering.
- Avoid stagnation of water in the orchard during the rainy season.

Flowering and Fruit Setting

- Flowering starts during the first week of March and flowering starts opening from the last week of March. Aonla is a cross-pollinated crop.
- During the dry period from April to July, fruits remain dormant in North India. Fruits start developing in the month of July, after the onset of rainfall. Initially, fruit development is slow and later develops at a faster rate in the month of September.
- Fruit matures during November - December.

Training and Pruning

Aonla plants should be encouraged to develop a medium-headed tree. The main branches should be allowed to grow at a height of 0.75 to 1 m above the ground level. Plants are trained in a modified centre leader system. Aonla trees do not require special training and running. However, during the initial years of planting, unwanted branches are removed during March- April.

Top-working

Older trees of inferior type can be converted into superior types by top working. Inferior trees are headed back to a height of 1.2 m from the ground during March and shoots sprouted on the stump are budded in July-August with the scion of the improved cultivar. In top working/frame working, more than one variety can be budded on a plant.

Intercropping

During the initial 4-5 years of pre-bearing stage, legumes and vegetables (short-duration crops) can be taken as intercrop. Karonda, phalsa can be taken as a filler crop.

Mulching and Weed Control

- Organic or inorganic mulch can be done to avoid evaporation losses of water and prevent weed growth.
- Application of diuron @ 3 kg/ha or glyphosate 3 l/ha during the second week of March in 500 litres of water to control weed flora in the aonla orchard.

Harvesting

When fruit attains maturity in the months of November- December, harvesting can be done and can be delayed up to February, but further delay may cause necrosis and decrease the crop of the ensuing season. Maturity can be judged by:

- The specific gravity of the fruit should be from 1.07 to 1.10.
- The colour of the fruit starts turning to light yellow at the maturity stage.
- The capsule colour changed from creamy white to brown-black.
- During harvesting, fruits should not be dropped on the ground.
- Seedling plants start yielding 10 years after planting, whereas budded/ grafted plants start bearing after 4-5 years. A mature tree yields approximately 200 kg of fruit/tree. The average yield is about 125 q/ha.
- The shelf life of aonla fruits at ambient temperature is about one week, but it can be extended up to 15 days by storing them in perforated polythene bags.

Insects

- **Bark-eating caterpillar (*Indarbela tetraonis*):** Caterpillar attack on stems and shoots of the growing trees. It feeds inside the hole during night and eats outside bark during day.

  Mix 10 ml monocrotophos 36% SL in 10 litre water and inject inside the hole during September-October. Inject Kerosene or any other insecticide inside the hole during February- March and plug it with cotton or mud.
- **Termite:** It attacks the roots and lower portion of the stem during dry period. In wild or unmanaged orchards, it reaches up to the top of the stem by making tunnels of soil.

  Apply chlorpyriphos @ 50 ml per plant and break the tunnels.
- **Shoot gall caterpillar:** It inserts into the shoot during the rainy season and forms a hard knot like structure on the shoot.

  Remove the infested shoots.

Diseases

- **Aonla rust:** The bluish-green pustules appear on the leaf and fruit surface.

  Spray Dithane Z-78 @ 0.2 per cent from July to September at 2-3 weeks intervals.

Physiological Disorder

Fruit Necrosis

- Turning of mesocarp to black and corky development of gummy pockets on the fruit surface.
- In this disorder, necrosis symptoms start with the browning of inner most part of the mesocarp tissues. It is mainly due to Boron deficiency.

Management:

- Grow resistant variety like Chakaiya.
- Control: Spray 0.6 per cent Borax during September – October at 15 day intervals. Grow Chakaiya, NA6 and NA7 cultivars almost free from this problem.
- Application of $ZnSO_4$ (0.4 %) + copper sulphate (0.4 %) and Borax (0.4 %) in September - October are effective in controlling internal necrosis.
- Francis variety is highly susceptible to this malady.

Chapter 8

Date Palm (*Phoenix dactylifera* L.)

Family: Arecaceae Palmaece, 2n =36

Introduction

- Date palm is a monocot plant with an unbranched fibrous stem is considered the as oldest among the cultivated fruit trees.
- It is considered to live in the desert because it can tolerate higher temperatures, drought and salinity than most of the other fruit plants.
- Fruits of date palm are the sweetest and most nutritive, and are good sources of iron, calcium, potassium and nicotinic acid. It is a good source of easily assimilable iron (7.3 mg/100 g) and hence prescribed to anaemic patients along with honey.
- Fully ripened dates (pind) contain 75-80 per cent sugars, whereas fresh dates (doka) have 60-65 per cent sugars. One kilogram of a fully ripe fresh date provides about 3150 calories.
- Dry dates are called Chhuhara, while fresh dates are called Khajoor.
- *P. sylvestris* (desi khajoor) is also widely spread in India and is an important source of jaggery (Khand) and a drink (palm wine) called tari/ toddy.
- Rajasthan and Gujarat (Kutch) are major date palm-growing states in India.

Origin

- It is believed to have originated in countries around the Persian Gulf of the middle east, southern Iraq (Mesopotamia) and Egypt.

Soil and Climate

- This crop can be grown on a wide range of soils, but deep loamy soils are suited the best.
- Soil should have better water-holding capacity with good drainage.
- It can be grown on saline and alkaline soils, but these affect crop growth, yield and productivity.
- The soil profile should be free from stone or hard pan of calcium carbonate up to 2 m.
- The pH of soil varying from 8.0 to 8.6 is suitable for the cultivation of date palms.
- Its successful cultivation requires moderate winter temperatures and long hot dry summers to mature fruits. It can withstand summer temperatures up to 50°C. The ideal temperature during flowering (February to April) and fruit ripening (May to August) is 25 to 40°C. The fruit development, maturity and ripening depend on the amount of heat unit accumulated or temperature during the fruit development phase. The crop is susceptible to frost.
- The heat unit requirement of date fruit varies from 1950 to 3650. It varies with cultivars.
- Rainfall, high humidity and cloudy weather hinder the ripening process of the fruits. The weather must be sunny and clear during ripening.
- A famous quotation (Arab saying) about its climatic requirement is "its feet in running water head in the fire of the sky".

Cultivars

- **Zaghloul:** Red-coloured fruit of a large size.
- **Halawy:** An early variety with small fruits less astringent at the doka stage.
- **Khadrawy:** It is a mid-season maturing dwarf cultivar.
- **Barhee:** Mid-season cv. with small to medium-sized yellow coloured fruits having low astringency at the doka stage. Suitable for raw eating (table purposes) while not fit for the soft date (khajoor).
- **Zahidi:** Late ripening cultivar with small to medium-sized fruits and heavy bearing. It is tolerant to rain and humidity, but fruits are astringent at a doka stage.
- **Medjool:** It is a late maturing cultivar; fruits are large-sized and suitable for dry fruits (Chhuhara). Highly astringency at the doka stage.
- **Shamran:** Mid-season, yellow at doka stage, faint longitudinal red streak near the base.
- **Hayani:** Less astringency at doka stage.

- **Khalas:** Early ripening, small to medium-sized, sweet at doka. Suitable for raw and soft dates.

Propagation

- **Seeds:** The long juvenile phase takes about 7-8 years to flower. The dioecious population is segregates into 50 per cent male and 50 per cent female flowers and sex can be determined at flowering. Seedling plants may not be uniform in bearing. This method is used mainly for breeding purposes. Germination of seeds is more than 85 per cent.
- **Sucker:** It arises from the lower base with a lot of roots. Generally, it should be 10-15 kg with roots.
- **Offshoots:** Arises above the ground level or on the stem. No root formation is there, so the survival rate is less. Trim the outer leaves to 2/3rd and inner leaves to ½ from offshoot at 4 to 5 days earlier to separation. Rooting in offshoots can be achieved by covering its attachment part with a cement bag or poly bag filled with rooting media. This is very tedious work and plants produce a limited number of offshoots or suckers in a life.
- **Tissue culture:** Produce true to type, disease free and healthy plants.

Planting

- The best time for maximum survival is August to September. Planting can be done during the spring season (February- March) also.
- Depending on the soil fertility, the spacing varies from 6 m to 7 m.
- For effective pollination, 5-10 per cent of male plants should be planted for commercial cultivation.
- Dig the proper pit (1 m^3) before planting and irrigate the plant after transplanting.
- Offshoots should be treated with Bavistin @ 0.1 per cent and chlorpyriphos 20 EC @ 0.1 per cent for better survival.

Manures and Fertilizers

- Apply 50 kg FYM per plant from September to December.
- Apply 600 g N (60%) + 100 g P_2O_5 + 70 g K_2O per plant three weeks before flowering.
- The remaining nitrogen (40%) should be applied during fruit development (March-April).

Irrigation

- The fruiting period of this crop is during the hot dry season of the year. So regular irrigation is required. This crop flourishes well under irrigated conditions.
- Prolonged water stagnation damages the plants.

- ☆ Apply frequent irrigation in young plants. Mulching is helpful in retaining moisture.
- ☆ Apply irrigation after fertilizer application.
- ☆ One irrigation is required before spathe formation for better flowering and fruiting.
- ☆ It requires regular irrigation during fruit development.
- ☆ Shortage of water during fruit development reduces the size of fruit, shrinks the fruit, causes premature drop of fruit, and reduces fruit weight and quality.
- ☆ Stop irrigation before harvesting which will be helpful to hasten the ripening.

Pollination

- ☆ It is a dioecious plant, so highly cross-pollinated. For more yield, pollinate the flowers 2-3 times during the season. It may give a metaxenia effect, so select the proper pollinizer.
- ☆ Under Indian conditions, the spathe (male flower) starts emerging in the month of February- March and the female flower starts opening in the month of March- April. The male flower spathe initiates early as compared to the female flower spathe.
- ☆ For storage, the pollen should be kept in sunlight for 6 hr, followed by keeping them in shade for 18 hr. After that, these pollens can be stored at normal temperature for 8 weeks in an airtight vial. However, under refrigerated conditions, these pollens can be stored for one year.
- ☆ Stigma remains receptive for several days.

Fruit Thinning

- ☆ To avoid excessive fruiting, remove some branches or strands from the bunches, or ½ bunch can be removed during pollination.
- ☆ It will be helpful in sufficient flowering and yield during the ensuing season.
- ☆ It also helps in early ripening of fruits or avoid delayed ripening.
- ☆ Thinning provides aeration in a bunch for proper development and ripening.
- ☆ The fruits may rot in compact bunches.
- ☆ Spray Ethrel @ 500-1000 ppm at the colour break stage to enhance early ripening.

Fruit Development Stages

Date palm fruits can be divided into four stages on the basis of fruit development and ripening:

- ☆ Gandora (Kimri): Hard green stage.
- ☆ Doka (khalal): Fruit becomes hard, yellow or pink or red, TSS between 30 to 45°B, astringency present or absent depending on cultivar and it is edible stage.
- ☆ Dang (rutab): Softness starts from the tip of the fruit; tannins and astringency disappear, lose weight and moisture content is about 35-40 per cent and it is an edible stage.
- ☆ Pind (tamar): Fully ripe fruit, lose weight, TSS between 60 to 84°B and edible stage.

Intercrop

- ☆ In areas with irrigation facilities, intercrop may be taken like lentil, gram, peas during winter and moong, urd during summer. Fruit plants of less height can also be taken.
- ☆ Additional quantity of irrigation and nutrients should be applied for intercrop.

Harvesting

- ☆ Depending upon weather conditions (mainly rainfall), fruit should be kept on the palm up to Tamar (pind) stage, if possible. Although fruits are edible after the Khalal (doka) stage, particularly in cvs. Halawy and Barhee. All the fruits in the bunch do not ripen at one time, so hand-picking of ripened fruits is required. The entire bunch can be harvested at 70 per cent fruit ripening.
- ☆ To enhance the ripening process or early ripening, spray ethrel @100 ppm on bunches at the colour break stage (*i.e.* greenish to yellow).
- ☆ Dry dates (Chhuhara): Fruits harvested at doka immersed in boiling water for 6-8 minutes and dried in an oven for 80-120 hours at 48-50°C or sun/solar drying for 10-15 days if the weather is dry. The product recovery will be about 45 per cent. Medjool is the best variety for dry dates.
- ☆ Soft dates: Fruit can be converted into soft dates by dipping the dung stage fruit into boiling water for 20-30 seconds, thereafter, dry them in the oven at 40°C or sun dry without any treatment.
- ☆ The drink of a date palm is known as dibbis.
- ☆ Date palm liquor 'Arrack' is popular in Iraq.

Storage

Fresh dates at Khalal/doka stage can be stored up to 2 months at 0 to 1°C and dry dates for 1 year at ambient temperature.

Insects

- **Termite:** Damage is more severe in newly planted orchards. It feeds on the root and lower portion of the stem by making tunnels of soil.

  Termites in young plantations can be controlled by the application of chlorpyriphos @ 30 to 50 ml per plant (as per age of plant).

  Rhinoceros beetle and red palm weevil: Adult beetles bore into unopened fronds and spathes. Irregular cut on the frond shoot is a symptom of attack.

  Control can be achieved by injecting monocrotophos @ 10 ml with 10 litres of water.

- **Black-headed caterpillar (*Opisina arenosella*):** It feeds on leaves and hide inside the tunnels in the folds of the leaves.

  Root feeding with 10 ml of monocrotophos mixed with 10 litre of water per tree.

Diseases

- **Fruit rot:** Fruit rot occurs due to heavy rainfall during the maturity of fruit due to high humidity.

  Provide better ventilation or avoid compactness of branches. Spray Indofil Z-78 @ 0.2 per cent.

- **Leaf smut (False smut):** During heavy rainfall, hard black smut-like pustules appear on the upper/lower surface of the leaflet and rachis. Formation of dark brown or black pustules spores on leaves (in humid conditions).

  Destroy affected leaves and spray 0.1 per cent Bavistin or 0.2 per cent copper oxychloride.

Chapter 9

Sapota (*Manilkara zapota* L. synonyms *Achras zapota* L., *Manilkara achras* (Mill.)

Family: Sapotaceae, 2n = 26

Introduction

- It is one of the important tropical crops.
- Sapota was first cultivated in Gholkward village in Maharashtra in 1898.
- India is the largest sapota producer in the world.
- Mainly grown for chikle *"gutta percha"* latex extracted from its stem.
- Used in the chewing and gum industry.
- In India, it is grown for sweet fruits. It is a good source of digestible sugar (12 to 18°B).
- It is cultivated in different countries in the world, mainly the West Indies, Philippines, Florida, Malaysia, Indonesia and Sri Lanka.
- The major sapota-producing states in India are Gujarat (27.8 %), Karnataka (26.8 %), Tamil Nadu, Maharashtra, Andhra Pradesh, West Bengal and Orissa.
- Cultivated successfully in subtropical areas of Haryana, Punjab and adjoining Uttar Pradesh.

Origin

- Sapota is native to Mexico and Tropical America, specifically Central and South America.

Soil and Climate

- This crop is grown on a variety of soils, however, deep, loamy, well-drained black soils are the best for better production and productivity.
- Trees will not thrive well in limey, sticky clay, hardy and water-logged soils.
- The problem of uprooting in sandy soils.
- Sapota is an important tropical region crop that needs a warm (11 to 34°C) and humid (>70 %) climate. However, temperature above 42°C in the summer causes drying of the stigma fluid of flowers and fruit drops.

Cultivars

- **Kalipatti:** Sweet in taste and grown in Gujarat, Karnataka and Maharashtra.
- **Cricket Ball:** Pulp granular and sweet, also called Calcutta large.
- **Chhatri:** Branches are drooping and high-yielding variety.
- **Culcutta Round:** Flesh is gritty and moderate quality, grown in Kerala.
- **Pala:** Fruits are small to medium and oval-shaped grown in Andhra Pradesh and Tamil Nadu.
- **Baramasi:** Fruits are medium-sized and round in shape and grown in northern India.
- **Bhuri or Bhuripatti:** Superior size fruit and plants are medium bearers.
- Other cultivars: Jingar, Vanjeet, Oval, Pot Sapota, CO 1, CO2, CO3.

  PKM1 (dwarf clonal selection from Guthi), PKM2, PKM3, PKM(Sa) 4, PKM(Sa)5.

Propagation

- Propagation of sapota is done by inarching/approach grafting in February-March, veneer grafting from July to September and soft wood grafting during July-August.
- **Air layering:** Select 1 to 2 years old branch of about 60 cm in length and 0.6 to 1.0 cm thickness during the onset of rains. Remove the bark in ring shape 2.5 to 3.0 cm in length and apply IBA @ 2000 ppm at the uppercut of the stem for better rooting. Cover the cut portion with moss grass and wrap it with a polythene sheet and tie it from both ends.

Rootstocks

Commercially propagated on seedling plants of rayan (Khirni- *Manilkara hexandra*) rootstocks.

Planting

The pits of 1 m^3 are dug in April-May at a distance of 8 to 9 m. Top 50 cm soil mixed with equal quantity of farmyard manure + 3 kg single super phosphate + 1.5 kg muriate of potash (MOP) and fill the pit.

- ☆ Planting should be done during July to September.
- ☆ High-density planting at 8 m x 4 m or 5 m x 5 m spacing improves yield.

Manures and Fertilizers

Table 9.1: Application of Manures and Fertilizers

Age of plant (years)	*Urea (g)*	*Single super phosphate (g)*	*Muriate of potash (g)*
1-3	100	125	125
4-6	200	250	250
7-10	400	500	500
11 and above	800	1000	750

- ☆ Apply a full dose of farmyard manure (FYM) 40 kg + single super phosphate (SSP)+ muriate of potash (MOP) in December – January.
- ☆ Apply ½ dose of urea in February and the remaining ½ in July-August, one meter away from the trunk, under the whole canopy.

Irrigation

- ☆ Younger plants should be irrigated at intervals of 6 to 12 days.
- ☆ Sapota can be grown under rainfed conditions, but for good yield, regular irrigation at fortnightly intervals during summer months and monthly intervals during winter months is required.
- ☆ Drought conditions or insufficient irrigation cause the dropping of flowers. Proper moisture can be maintained through drip irrigation with limited water availability. It saves about 40 per cent of water and increases net income by 70-75 per cent.

Flowering

- ☆ Under tropical conditions, sapota remains evergreen throughout the year. Peak flowering seasons are February to March and July to November, fruits are harvested during September-November and January –March, respectively.
- ☆ Flowering takes place on the previous season's shoots.
- ☆ It is a cross-pollinated crop due to its protogynous flowers. Therefore, planting of single cultivar is not recommended.

- The fruit setting is very high, but fruit retention is very poor due to heavy fruit drop in early stages. A spray of NAA @ 300 ppm at the pea stage is effective in enhancing yield and quality.
- Sapota fruit is drupe. The growth of fruits follows a double sigmoid growth curve.

Harvesting

- Sapota may start yielding from the 2nd to 3rd year of planting. However, full yield starts from the 7th year of planting.
- Time taken from flowering to maturity varies from variety to variety, but in most cases, it takes about 7 to 10 months.
- It is a climacteric fruit and it is very difficult to determine the physiological maturity of the fruit. Fruit changes to dull orange or potato colour and reduced brown scaly material from the fruit surface. Stigma at the end of the fruit is dried and detached easily.
- Fruits are graded according to size. For uniform and rapid ripening, fruits should be dipped in 1000 ppm ethephon for 2 minutes.
- Sapota fruits can be stored for 5 to 6 weeks at low temperature (2 to 3°C) with relative humidity of 85 to 90 per cent.

Insects

- **Stem borer (*Isocrata tetraonis*):** The larval stage makes a hole in the stem and feed inside the hole.

  Drench the hole with Kerosene oil @ 0.1 per cent and plug this hole with mud/cotton.
- **Leaf miner (*Acrocercops gemoniella*):** It attacks mainly during the rainy season. The maggot feeds inside the leaves and leaves become white and dry during severe damage.

  Spray once or twice with dimethoate (0.05 %) or malathion (0.05 %).
- **Mealybug (*Phenacoccus icerjoides*):** Both nymph and adult suck the sap and feed on leaves and fruits in colonies. It looks like a cottony appearance.

  Spray dimethoate @ 0.05 per cent.
- **Scale insects (*Pulvinaria psidii*):** Both nymph and adult suck the sap and female gives birth to 200-300 nymphs.

  Spray dimethoate or malathion @ 0.05 per cent.
- **Stone borer:** The adult lays eggs on immature fruits and leaves and feeds on endosperm.

  Spray malathion @ 0.05 per cent or spray fenvalerate @ 0.01 per cent.

Diseases

- **Leaf spot (*Phaeoleospora indica*):** A pink to brown spot appears on mature leaves with a whitish centre.

  Grow varieties resistant to leaf spot, *e.g.* Cricket Ball and Co 1, or tolerant to leaf spot, *e.g.* Kalipatti and Co 2. Spray Dithane Z-78 @ 0.2 per cent at monthly intervals. Avoid to grow susceptible varieties *e.g.* Calcutta Round.

- **Sooty mould:** Fungus develop on the excreta of mealybug and scale insects and honeydew like substances form on leaves, it hinders the photosynthesis and translocation of food.

  Spray 0.2 per cent Zineb or spray starch solution @ 0.5 per cent. Starch forms the flakes and drops off.

Chapter 10

Plums (*Prunus* spp.)

Family: Rosaceae; 2n= 36

Introduction

- Low-chilling cultivars can be grown in sub-tropical areas of Punjab, Haryana and Uttar Pradesh.
- In India, its production is 84450 MT in an area of 23510 ha.
- At the global level, India ranks 7th in the area and production of plum.
- Plum is rich in minerals, vitamins, sugars, organic acids, protein, fat, Ca, Mg, K and Fe.
- Fruits are used to prepare different products such as jam, jelly, juice, nectar and marmalades. High sugar-containing plums can be dehydrated. The dried one can be used for the preparation of chutney.
- High sugar and sorbitol containing varieties are used for the preparation of wine and brandy.
- Seeds of plum have about 40-50 per cent oil, which is used for cosmetic and medicinal purposes. 'Prune' can be dried successfully with or without stone.
- *Prunus saliciana* is an important temperate fruit crop commercially grown in the north-western Himalayas, *i.e.* Uttarakhand (41.25 %), Jammu & Kashmir, Himachal Pradesh.

Origin

- *Prunus salicina* (Japanese Plum): Originated in south-western China but is more popular in India.

- *Prunus domestica* (European plum): Origin in the Caspian Sea.
- *P. americane* (American Plum): Origin in North America.
- *P. cerasifera* (Cherry plum): Origin in Europe, West Asia.

Soil and Climate

- It should be grown in deep, well-drained, fertile sandy loam soil. The pH between 5.5 to 6.5 is most suitable. Heavy, poorly drained, water-logged, salty soils with hard pans should be avoided.
- Japanese plums require less winter chilling, *i.e.* 700-1000 hrs below 7.2°C, whereas European plums require 1000-1200 chilling hours. Plum blooms early in the season, thus, they are prone to frost injury. An area with frost-free spring, having proper water drainage and adequate sunshine in summer is most suited for plum cultivation. Plum orchards are prone to damage by high wind velocity and must be protected by planting windbreak trees.
- Japanese plum is sensitive to early summer frost. If there is no frost or protected from frost injury, it bears good yield.

Cultivars

- **Satluj Purple (Tiwan x Methyley):** The fruit's colour is bright crimson at maturity with orange-yellow flesh. Fruit is thick-skinned, medium-sized and round in shape.
- **Titron:** It is a self-fruitful variety, but yield improved with pollinizer Alucha Early Round. The fruit is round in shape and the skin is deep purple coloured. Fruit is suitable for jam-making.
- **Alu Bokhara:** It is self-fruitful, but for better fruiting, it needs pollinizer Hawe. The fruit is large in size and the skin has red tints.
- **Kala Amritsari:** It is self-fruitful and good for plain areas. It performs better with pollinizer Titron. Both ends of the fruits are depressed. Fruit skin is brown in colour and good for jam-making.
- **Santa Rosa:** It is a self-fruitful variety. Fruits are large-sized and heart-shaped. Fruit skin is purple and crimson in colour.

Low Chilling Variety

Titron, Alucha Black, Alubukhara, Kala Amritsari, Peshawari Kala, Alucha Yellow.

Pollinizers

Beauty, Santa Rosa and Mariposa are self-fruitful cultivars (about 30 % fruit set) of plum and are used as pollinizers.

Propagation

- Cuttings (hardwood) should be treated with IBA @ 2500 ppm for 10 seconds during December-January.
- Tongue grafting during February.
- T-budding (shield budding) during April and July-August.

Rootstock

- **Wild apricot (Zardalu)** *Prunus armeniaca.*
- **Kabulgreen:** Gauge plum in Punjab is used for heavy and wet soil.
- **Myrobalan B:** It is multiplied through mound layering (stooling) for clonal rootstock. It is resistant to bacterial gummosis.
- **Myrobalan 27:** Drought tolerant.
- **Myrobalan 29C:** Tolerant to high soil moisture conditions.
- **Pixie, St. Julien 1,2,3:** Gave dwarfing effect on the scion.
- **Myrabi:** Semi-dwarf, tolerant to water logging.
- **Marianna 2624:** Resistant to nematode, crown gall, cold hardy and tolerant to high soil moisture.
- **Peach seedlings** are recommended for sandy loam soils.

Planting

Planting at 6 m x 6 m in December – January and July-August. Plants of Kala Amritsari are also planted in orchards of Satluj Purple as pollinizers in a 15: 85 ratio.

High-density planting

High-density planting can be done with Satluj Purple plum. It can be planted at a distance of 6.0 m x 1.5 m when trained in a modified centre leader system and 10-15 cm pinch during summer. This will provide good yield as well as better quality fruits.

Manures and Fertilizers

Table 10.1: Application of Manures and Fertilizers

Age of plant (years)	*Farmyard manure (kg)*	*Urea (g)*	*Single super phosphate (g)*	*Muriate of potash (g)*
1	6	60	100	60
2	12	120	200	120
3	18	180	300	180
4	24	240	400	240
5	30	300	500	300
6 and >	36	360	600	360

- ✰ Basal doses of fertilizers such as farmyard manure (FYM), single super phosphate (SSP) and muriate of potash (MOP) should be applied in the months of December- January.
- ✰ The half-dose of urea should be applied before flowering and the remaining half after fruit set.
- ✰ To recover the plant from Zn deficiency, spray $ZnSO_4$ @ 0.5 per cent along with unslaked lime in sandy soils.

Irrigation

- ✰ Require regular irrigation. Peak water requirement is from April to June, which coincides with fruit development. About 6 irrigations are sufficient in mid-hills.
- ✰ In water scarcity areas, mulching can conserve moisture. Mulching and drip irrigation can be adopted for water saving. Avoid irrigation at flowering.

Training

- ✰ Modified leader system in a subtropical region.
- ✰ Open centre system in temperate region.

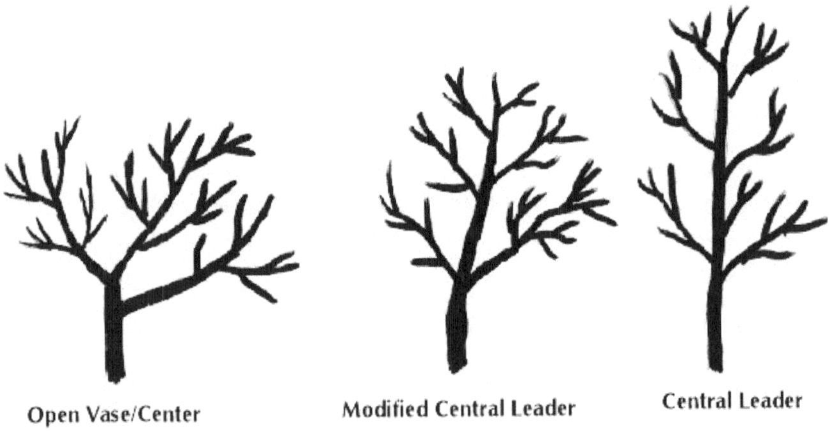

Fig. 10.1: Common Training Systems in Fruits

Pruning

Plum is considered spur bearing, but at the same time, some fruits are produced on 1-year old lateral shoots. Heavy pruning or heading back produced more water shoots, which results in poor or no fruiting. So annual pruning should be done to remove the dried, dead, diseased, crisscross branches, water sprouts and suckers.

Pruning intensity differs with varieties:

Japanese plums are pruned heavier for good bearing, while European plums are pruned lighter to renew old spurs.

Pollination and Fruiting

Self-fruitful is not a constantly stable character and may vary from year to year and area to area. Honeybees are major pollinators. One bee colony per acre improves fruit set. The cultivar Santa Rosa is self-fruitful in India and in some areas of the USA, whereas it a gives heavy yield when planted in solid blocks. Every 3rd plant in the 3rd row should be a pollinizer plant.

Fruit Thinning

Spray 200 ppm ethephon at full bloom stage for thinning small-sized fruits of poor quality. Thinning of fruits (excessive bearing) is helpful to improve the quality and colour of fruits and avoid breaking of limbs.

Harvesting

Maturity Indices

1. Colour development (skin becomes coloured).
2. Flesh firmness (15 ± 2N).
3. Days after full bloom, *e.g.* Beauty Frontier takes 87 days and Alu Bukhara 102 days.
4. TSS: 12.5°B for European plum.
5. The TSS/acid ratio is 12 to 15.

Plants start bearing in 2-3 years after planting. Harvest the fruits during May-June in the subtropical and June-July in the hills. For distinct markets, fruits should be harvested at a 50 per cent colour development stage. Follow proper grading and packaging before marketing.

Grading

Plums are graded on the basis of size of fruits in three categories *viz.*, Special grade size above 42 mm, Grade I having size between 36-42 mm and Grade II having size less than 36 mm.

Storage

The fruits of plum are perishable in nature. These fruits can be stored for 2-4 weeks at low temperatures (0°C) with 85-90 per cent relative humidity. Dipping of fruits in 4 per cent $CaCl_2$ for two minutes increases shelf life for 12 days at room temperature.

Insects

1. **Leaf curl aphid:** Nymph and adult suck the cell sap, leaf becomes curled and drops down in severe cases. It also affects the flowering and fruiting of plants.

   Introduce *Aphidius colemani* as a parasite of this aphid. Important predators include lady beetles, soldier beetles, green lacewings, brown lacewings and syrphid flies.

2. **Fruit fly:** The female lays eggs on the mature fruit, larvae/ maggot feed pulp and fruit becomes unfit for consumption.

   Use a pheromone trap to attract the male flies.

Diseases

1. **Shot hole:** Small purple-black lesions appear on the leaves. These spots enlarge with the light centre and purple-brown margin, and the leaf may drop out.

   Foliar application of Bordeaux mixture or copper oxychloride @ 3 g per litre water.

2. **Brown rot:** It infects shoot blossom and fruit. The Infected blossom turns brown and drops down.

   Foliar application of copper oxychloride @ 3 g per litre water.

Chapter 11

Citrus (*Citrus* spp.)

Family: Rutaceae, 2n=18

Introduction

- ☆ Nagpur mandarin was introduced in India in 1894 by Shuji Raja Bhonsle.
- ☆ ICAR-Central Citrus Research Institute, Nagpur, Maharashtra.
- ☆ In the citrus group, mandarins are the most important, followed by sweet orange and limes/lemons.
- ☆ At the global level, India ranks 3rd in area and production of citrus.
- ☆ The citrus fruit is called hesperidium.
- ☆ Seeds are monoembryonic as well as polyembryonic.
- ☆ Citrus is a micro-nutrient loving plant.
- ☆ "Limolin": The glycoside in citrus fruit is responsible for the bitter taste of juice.
- ☆ Mandarins and sweet oranges are highly susceptible to water logging.
- ☆ "Troyer Citrange" rootstock (1.8 m x 1.8 m) used for high-density planting of Kinnow. Ultra-dwarf rootstock in citrus is Flying Dragon.
- ☆ Classification of citrus was given by Tanaka and Swingle.
- ☆ Citron (*C. medica*) is also known as the Persian apple.
- ☆ The United States is the largest producer of citrus, with 35-40 per cent of world production.
- ☆ Major citrus-growing states in India are Maharashtra, Andhra Pradesh, Punjab, Karnataka and the North-Eastern region.

Origin

- South East Asia, tropical and sub-tropical regions of Indo-China.

Soil and Climate

- Citrus trees are found/grown between 40° N and 40°S latitudes in all countries.
- Trees thrive well on deep, well-drained, loamy and fertile soils.
- Soil pH between 5.5-7.5 and salt concentration of irrigation water <1000 ppm, EC <0.8 dSm^{-1} are suitable for the citrus group.
- Poor drainage and high-water tables are unfavourable.
- High concentrations of chloride ions show a detrimental effect on growth.
- Perform well or produce good quality fruits in dry, semi-arid climates.
- It can be grown at temperature between 14 to 40°C and for best growth, it should be around 29 to 35°C.
- Vigorous roots and shoot growth occur between 23 to 27°C soil temperature and growth ceases below 10°C (soil temperature).
- It can tolerate occasional light frosts.
- Grapefruit can tolerate hot arid conditions up to 48.8°C. On the other hand, lime and lemon show symptoms of hot injury at 35 to 37.8°C.
- Fluctuation in day and night temperature helps in better colour development, accumulation of more sugar and formation of acids.
- Heavy rains at the time of flowering and fruit setting are unfavourable.

Citrus Classification

The classification of citrus was given by two famous authorities, W.T. Swingle (USA) and T. Tanaka (Japan) are at two extremes.

- Swingle (1948) recognized only 16 species under the genus *Citrus* whereas Tanaka (1954) described as many as 144 species.
- Swingle divided the genus *Citrus* into 2 subgenera *viz*; *Eucitrus* (10 species) and *Papeda* (6 species).
- Tanaka divided the genus *Citrus* into 2 subgenera *viz.*; *Archicitrus* (98 species) and *Metacitrus* (46 species).

Major Species of Horticultural Importance

1. Mandarin Group

- It occupies the first position among citrus in India in area and production.
- Seeds have polyembryonic properties.
- Trees are medium sized with upright growth.

- ☆ Fruits are medium-sized and loose-skinned.
- ☆ Major mandarin producing states in India are Madhya Pradesh (41.2 %), Punjab (23.7 %), Maharashtra, Rajasthan, Assam *etc.*
- ☆ **Sangtra (*Citrus reticulata*)**: Chinese origin. Polyembryonic cultivars are Nagpur, Coorg, Khasi in India; and Ponkan in China.
- ☆ **Kinnow:** *C. nobilis* (King) x *C. deliciosa* (Willow leaf) was given by H.B. Frost in 1935 and it reached India in 1959.
- ☆ NRCC Nagpur Mandarin Seedless 4, PAU Kinnow 1.

2. Orange Group

- ☆ *C. sinensis* (Sweet orange or Malta): Polyembryonic cultivars are Mosambi, Jaffa, Hamlin, Blood Red, Sathgudi, Valencia, Pineapple, Washington Navel Orange, Pusa Round, Pusa Sharad.
- ☆ It is the second most important crop in the citrus group.
- ☆ India ranks 1st in area and 3rd in production in the world.
- ☆ Fruits are tight-skinned.
- ☆ The leaf apex is blunt and pointed with narrow winged petioles.
- ☆ Mosambi is an important member of this group. Its apex is marked with a circular ring.
- ☆ Pineapple and Valencia are indicators of greening.
- ☆ The major orange producing states in India are Andhra Pradesh (61 %). Maharashtra (21 %), Telangana, Madhya Pradesh, Karnataka and Punjab *etc.*

3. Acid Lime Group (Lime, Lemon and Citron)

- ☆ ***C. limon* (Lemon):** Weakly polyembryonic cultivars are Eureka, Baramasi, Lucknow Seedless, Pant Lemon 1, Kagzi Kalan, Baramasi and Galgal. Leaves are narrowly winged, the tip is sharply pointed with a sub-serrate margin.
- ☆ India ranks first in production (16.3 %) and area of lime and lemon.
- ☆ The major lime and lemon producing states in India are Gujarat (19.2 %), Andhra Pradesh (17.8 %), Madhya Pradesh, Karnataka, Orissa, Maharashtra *etc.*
- ☆ ***C. jambhiri,* Rough lemon *(Jatti Khatti)*:** Polyembryonic Indian origin, popular rootstock, fairly tolerant to viral diseases.
- ☆ ***C. aurantifolia* (Acid lime or Sour lime):** Polyembryonic, popular cultivars are Kagzi Lime, PKM-1, Pusa Abhinav, Pusa Udit, Thar Vaibhav, Sai Sarbati. Sai Sarbati is susceptible to tristeza virus and canker. Trees are bushy and leaves are small in size with narrow winged petioles. Fruits are small in size and thin skinned available throughout the year.

- ☆ ***C. karna* (Kharna Khatta):** Popular rootstock.
- ☆ ***C. limonia* (Rangpur lime):** Hardy and popular rootstock, tolerant to tristeza virus and salt. It is the most promising rootstock for mandarin.
- ☆ ***C limettioides* (Sweet lime):** Medium-sized tree, leaves are medium-sized, pale green, margin is serrated and have a big thorn. Sweet lime is resistant to greening.

4. **Grapefruit and Pomelo Group**

- ☆ ***C. paradisi* (Grapefruit):** Origin-South China. It is polyembryonic and also called forbidden fruit and breakfast fruit. Cultivars are Foster, Star Ruby (developed through mutation breeding from Hudson), Marsh Seedless, Duncan, Red Blush, Triumph.
- ☆ ***C. grandis* (Pomelo):** It is monoembryonic, origin is Malaysia and Polynesia. Leaves are pubescent on the lower surface. Cultivars are Kao Pan, Arka Anantha, NRCC Pummelo 5

Related Genera

- ☆ ***Poncirus trifoliata* (Trifoliate orange):** Fruits are inedible, polyembryonic and used as rootstock. Resistant to *Phytophthora* and nematode.
- ☆ ***Fortunella* sp. (Kumquat):** Polyembryonic, ornamental with small oval fruits.

Intergeneric Hybrids

- ☆ Citrange: *Poncirus trifoliata* (Trifoliate orange) x *C. sinensis* (Sweet orange). Cultivars are Troyer, Carriyo the Stonia.
- ☆ Citrumelo: *Poncirus trifoliata* (Trifoliate orange) x *C. paradisi* (Grapefruit).

Intrageneric Hybrids

- ☆ Tangor: *C. reticulata* (Mandarin orange) x *C. sinensis* (Sweet orange).
- ☆ Tangelo: *C. reticulata* (Mandarin orange) x *C. paradisi* (Grapefruit).

Propagation

Citrus plants are propagated by budding (T or Patch) during August-September or February-March. The cutting/layering methods of propagation are followed in lime and lemon.

Important Rootstocks

- ☆ **Rough lemon (*C. jambhiri*):** This is considered the best rootstock. It is resistant to the tristeza virus and tolerant to frost. It is better for sandy and infertile soils. Slightly tolerant to saline and calcareous soils.
- ☆ **Rangpur lime (*C. limonia*):** It is a popular rootstock for mandarins and sweet oranges. Adaptable to various soil conditions and salt tolerant.

- **Cleopatra mandarin (*C. reshni*)**: This rootstock is better for Malta and grapefruit. Its response to heavy bearing and tolerance to wilt and other diseases.
- **Karna khatta (*C. karna*):** The plant grows vigorously. Silty and clayey soils are better for stock.
- **Sour orange (*C. aurantium*):** Resistant to nematode and frost conditions but susceptible to tristeza virus and gummosis.
- **Trifoliate orange (*Poncirus trifoliata*):** It is resistant to cold, *Phytophthora* and nematodes, it produces a more dwarfing effect than other rootstocks.
- **Troyer citrange and Flying Dragon:** Dwarfing rootstock.

Planting

- Dig a pit of 1 m^3 and fill it with 50 per cent topsoil + 50 kg of farmyard manure + 2 kg single super phosphate.
- Planting is normally done during the monsoon season or spring season.
- The appropriate spacing for planting is 6 m x 6 m.
- Kinnow and Baramasi lemon plants can be successfully transplanted bare rooted from December to February.

Manures and Fertilizers

- Apply fertilizers on the basis of soil testing or leaf nutrient analysis.
- Fertilizers should be applied 30-40 cm away from the trunk below the canopy of the tree. Irrigate the plants just after the application of manures and fertilizers.

Table 11.1: Manures and Fertilizers

Age of plant (years)	*Farmyard manure (kg/plant)*	*Urea (kg/plant)*	*Single super phosphate (kg/plant)*	*Muriate of potash (kg/plant)*
1-3	10-40	0.200-0.600	0.250-0.750	0.100
4-6	40-70	0.750-1.250	1.000-1.500	0.150
>7	100	1.500	2.000	0.175

- Apply farmyard manure, single super phosphate, muriate of potash (MOP) in the last week of December and urea in two split doses, first ½ in mid-February and remaining ½ in April.
- In Haryana, most of the orchards show a deficiency of Zn and nitrogen. Zn deficiency in sweet oranges causes chlorosis of leaves, commonly known as mottled leaves (terminal leaves become small and narrow). To overcome this deficiency, apply 0.5 per cent zinc sulphate and 1.0

per cent urea through foliar application in the months of May-June and August- September.

Irrigation

- Water requirement in citrus varies with the types of soils and weather conditions.
- The fruit development stage is the most critical stage for irrigation.
- Water should be applied under the whole canopy of the plant but away from the main trunk.
- The modified basin method is suitable for the application of water to the plant. However, now-a-days, drip irrigation become more popular. This method has the highest water use efficiency.
- Irrigation should be applied before sprouting in February and after fruit set in April and during hot weather, otherwise the growth of the tree may be adversely affected, which results in excessive shedding of flowers/ fruits.

Flowering

- The flowering of citrus takes place mostly in spring.
- In lime and lemon, flowering takes place throughout the year.
- Flowering can be regulated by low-temperature stress, moisture stress (withhold irrigation), balance of C:N ratio, growth retardant (paclobutrazol) and adjustment of fruit harvesting.

Fruit Thinning

- The Kinnow mandarin is capable of producing heavy crops. During the initial years, heavy crops not only impaired the tree's health but also reduced the fruit size and sometimes made it unable to set fruit in the ensuing season.
- Thinning of the fruits and flower buds should be done to keep a balance of a tree's foliage and the number of fruits per tree.

Harvesting

- Citrus fruits are non-climacteric in nature, so these fruits should be harvested at appropriate maturity indices or stages.
- In North India, fruits are harvested during November-February, whereas in South India, fruits are harvested during October-March.
- Post-harvest handling includes washing, drying, sorting, grading, wrapping in tissue paper and packaging in corrugated boxes.
- Fruits dipped in 500 ppm Benlate or 0.1 per cent carbendazim (Bavistin) can be stored for 20 days at room temperature.

- ☆ Malta fruits can be stored for 2 to 3 months at 4.4°C with 85 to 90 per cent relative humidity.
- ☆ Mandarins can be stored for many months at 8 to 10°C with 85 to 90 per cent relative humidity. However, under room temperature, they can be stored well for 3 to 4 weeks.
- ☆ Acid limes can be stored for 6 to 8 weeks at 8.3 to 10°C and 85 to 90 per cent relative humidity.
- ☆ Lemons can be stored for 8 to 12 weeks at 7.2 to 8.6°C with 85 to 90 per cent relative humidity.

Insects

Citrus psylla: It is more active during new growth flush, *i.e.* March- April, June-July and September – October. Nymphs and adults suck the sap from growing shoots.

Apply dimethoate (Rogor) @ 750 ml or oxydemeton methyl (Metasystox) 750 ml in 500-liter water. Avoid spray at the time of flowering and pollination.

Citrus leaf miner: It feeds on leaves by creating shallow tunnels and mines. Control is the same as above.

Lemon butterfly: Larvae feed on the leaf from margin to mid rib. During severe attacks, it defoliates the whole plant.

Spray monocrotophos @ 500 ml in 500-liter water.

Citrus whitefly: It sucks the sap from leaves. Honeydew like substances secreted by whitefly become blackish on leaf and fruit surfaces.

Spray monocrotophos @ 500 ml in 500-liter water.

Citrus black fly: It sucks the sap from leaves and excretes honeydew like substance which converts the leaf into black colour and later on, premature leaves fall.

Spray monocrotophos @ 500 ml in 500-liter water.

Termite: Moisture and cellulose are required for the survival of termites. Keep the orchard clean, deep ploughing and avoid application of un-rotten manure and remove farm residue.

Apply chlorpyriphos 50 ml per plant or 1 litre per acre, along with irrigation.

Bark-eating caterpillar: Adult bore into the trunk near the junction of branches. It feeds inside the hole during the day and outside bark during the night.

Inject a solution of monocrotophos @ 10 ml per 10-litre water and drench it in the hole of the insect, or drench Kerosene oil 10 per cent + 100 g soap + 9-litre water in the insect hole.

Diseases

- **Canker:** Dark brown, rough, raised areas on leaves, twigs and fruits.
- **Gummosis or foot rot (*Phytophthora palmivora*):** Death of bark at soil line (collar region) on the trunk, killing of wood and gumming stem.
- **Mottle leaf (Zinc deficiency):** Chlorotic areas in between the main and lateral veins on each side of the midrib.
- **Stem and fruit rot:** Dark circular depressions with yellowish margins on leaves, branches and fruit; later spots become raised, rough and light brown.

  To control the above diseases follow the following spray schedule:
- December to February: i) Treatment of foot rot/gummosis by decortications and disinfection of wounds on the trunks with Bordeaux paste.

  ii) After pruning, spray 0.3 per cent copper oxychloride in October, December and February.
- During April –May, spray 0.3 per cent copper oxychloride.
- Apply 3 kg zinc sulphate + 1.5 kg lime in 500-litre of water to meet the zinc deficiency.
- After the first shower of rainfall (during July), spray 0.3 per cent copper oxychloride.
- August-September: For canker disease of Santra, Malta, repeat spray of 0.3 per cent copper oxychloride and also repeat spray of zinc sulphate + lime mixture.

Citrus Nematode

Citrus nematode (*Tylenchulus semipenetrans*) is one of the factors of citrus decline. The leaves and branches start dying from the upper side/ tip and it increase gradually towards the lower side. On weak plants, there is a development of small fruit and fruit starts dropping before maturity. The root of the plant becomes deformed and the soil particles remain stacked with the root and turn in a dull colour. In case of a severe attack, the root beneath the bark becomes black and muddy spots appear.

Planting material should be nematode-free. Use resistant rootstock. Do not allow the flow of water from infected plants to healthy plants and use neem-based botanical pesticides.

Physiological Disorder

i) Fruit Drop

Drops of flowers and fruits at different stages are a natural means to adjust its resources.

Period of fruit drop

Pin head fruit drop: This occurs primarily due to unfertilized ovaries. These are normal and may not affect the yield.

June drop: The second wave of fruit drop in citrus comes at the onset of hot weather. At this stage, fruit of 1-2 cm diameter may drop. The prevalence of high temperature, low humidity and strong winds create conditions of moisture stress in the tree. Consequently, a water deficit occurs, which leads to excessive shedding of young fruits.

Premature fruit drop: In India, premature fruit drop is a serious problem in many places and it is often responsible for causing heavy losses to fruit growers. The fruit can drop prematurely owing to several reasons. Some fruits may drop down due to mechanical injury, splitting, attack of pests and disease and sometimes without any visible symptoms.

Causes of Fruit Drop

(a) Physiological fruit drop: Pre-harvest drop of well-grown fruits prior to maturity is a serious problem for reducing the yield. The main causes may be:

i. Climatic factors
ii. Improper water management
iii. Lack of nutrition
iv. Seed per fruit.

Symptom: The flowers as well as fruits drop is primarily due to the formation of an abscission layer at the point of attachment of the fruit with the twig. The balance of growth regulator within the tree is responsible for the formation of the abscission layer.

Control: Two sprays of 10 ppm 2,4-D + 0.5 per cent $ZnSO_4$ + 20 ppm aureofungin in the month of June-July and the second week of September. In cotton growing areas replace 2,4-D with NAA @ 20 ppm.

(b) Pathological fruit drop: Fruit drop may be due to attack of insect pests and diseases.

Causes: i. Stylar end rot ii. Stalk end rot

Control: Apply Ziram (1250 g) or Bavistin (500 g) in 500-litre of water/acre during mid-April, August and September.

(c) Mechanical fruit drop: Fruit drop due to high wind velocity and other mechanical injury.

Control: Plants of trees on the boundary as the wind breaks.

ii) Granulation

Reported by Bartholomew in 1934 from California. It is also known as corkiness/crystallization/dry end/kaosarn/sclerocystis. Granulation is a complex pre-harvest disorder of citrus fruits characterized by the apparent drying up of juice vesicles, which become hard, assume a grayish colour and are enlarged with an increase in pectin, lignin and other polysaccharides results in a considerable decrease in total soluble solids, acidity, juice percentage and sugars. It occurs more in young vigorous trees than old trees. Sweet oranges are more prone to this disorder than mandarins. The causes are as under:

i) High relative humidity.
ii) High temperature during fruit development.
iii) Apply frequent irrigation.
iv) High vigour of plant.
v) Large fruit size.
vi) Heavy crop load on the plant
vii) Application of high nitrogenous fertilizers.
viii) Use vigorous rootstocks like Rough Lemon, Karna Khatta, Trifoliate orange and Sour orange.
ix) Sweet orange cultivars Hamlin, Mosambi and Mandarin cultivar Dancy Tangerine are highly susceptible.

Factors affecting granulation

- ☆ **Climate**: Humid climate is reported to have a higher incidence of granulation than trees growing in dry regions.
- ☆ **Species and cultivars**: Sweet oranges are more prone to granulation than mandarins, grapefruit and tangelos.
- ☆ **Rootstocks**: The vigorous rootstock, like rough lemon, induced a higher incidence of granulation. Blood Red on Cleopatra and Pineapple on Troyer exhibit the least granulation.
- ☆ **Mineral nutrition**: Deficiency of Zn enhances lignin formation.
- ☆ **Enzymes and plant growth regulators**: Low auxin and increase in pectin content increases granulation.
- ☆ **Crop load**: Heavy crop load increase the incidence.
- ☆ **Tree location**: More granulation was noticed in trees planted in the interior of the orchard, interior part of the tree canopy and northern side of the tree.

Control: Following can be tried as remedies, but no one is a successful method.

- ☆ Decrease the frequency of irrigation.
- ☆ Avoid excessive use of nitrogenous fertilizers.
- ☆ Avoid the use of vigorous rootstock.
- ☆ Early harvesting of crops.
- ☆ Spraying lime, $ZnSO_4$, Bordeaux mixture or $ZnSO_4$ + $CuSO_4$ (0.5 %).
- ☆ Supply nutrients like Zn, Cu, K each at 0.25 per cent commencing from August till September during fruit development.

iii) Citrus Decline

- Citrus decline may be due to unhealthy planting material carrying viruses, diseases, improper rootstock, mismanagement of orchards, defective soil, poor drainage, malnutrition, insect-pest, nematode and diseases.
- Tristeza, a very serious viral disease causing various deficiency symptoms like leaf fall, root decay, dieback of shoots and finally, death of the tree.
- Trees normally produce heavy crops before dying.

Control: Though appropriate control measures are not available as a means of rejuvenating the declining plants, the following measures may be followed.

- Remove all dead wood before new growth starts and spray Bordeaux mixture and cover all cut surfaces with Bordeaux paste.
- Apply more organic manures and nitrogen as per recommendation.
- Regularly follow the recommended plant protection measures and spray 0.3 per cent zinc sulphate and Bordeaux mixture during April, June and September.
- New plantations should be done using Rangpur lime as rootstock, which is reported to be highly tolerant to tristeza.

Chapter 12

Grape (*Vitis vinifera* L.)

Family: Vitaceae; 2n=38

Introduction

- Grape is a temperate fruit crop but has been adopted successfully in subtropical areas.
- The grape is believed to have been introduced in India by the invaders from Iran and Afghanistan about 1300 A.D.
- Grape is one of the important fruit crops of India. India ranks 4th in average productivity (21 t/ha) in the world.
- The genus *Vitis* is sub-divided into two sub-genera: *Muscadinia* (2n=40) and *Euvitis* (2n = 38).
- *Vitis vinifera* is the most popular and commercially grown deciduous climber plant with forked tendrils and shiny leaves.
- Harold Olmo (HD Olmo) is a Grape Breeder, Professor, University of California.
- Tartaric acid is commercially extracted from grapes.
- The peninsular portion of India is a major grape growing area (90 %). The grape-growing states are Maharashtra (78 %), Karnataka (18 %), Tamil Nadu, Andhra Pradesh, Mizoram and the north western region.

Grapes are used for the following purposes:

1. Table purpose: Most of the varieties grown in India are used for table purposes. Major table purposes cultivars are Cardinal, Muscat Hamburg, Perlette, Anab-e-Shahi, Tokay, Thompson Seedless, Concord, Pusa Seedless, Red Malaga, Muscat of

Alexandria, Delware, Catawba, Ohanez, Emperor and Italia etc. These are used for fresh consumption with the following qualities.

- ☆ Good eating and shipping quality.
- ☆ Attractive in shape, size and colour.
- ☆ Good keeping quality.

2. Raisin making: These are consumed as dried grapes commercial raisin making cultivars are Thompson Seedless, Red Corinth, Black Monukka, Cape Currant, Seedless Sultana and Kishmish Beli, which should have the following characteristics:

- ☆ Soft texture.
- ☆ It has a good sugar content.
- ☆ Pleasant flavour.
- ☆ Non-sticky after drying.
- ☆ Both small and large sized.
- ☆ Seedless.

3. Juice making: The cultivars have a good amount of acceptable juice content.

- ☆ Retain good and natural flavour.
- ☆ Acceptable for beverages.
- ☆ Clarification in juice and can be preserved for a long period.
- ☆ Concord grapes are used in the United States of America.
- ☆ White Riesling and Chasselas Dore are used in Central Europe.
- ☆ Aramonand Carignan is used in France.

4. Wine making and canning: The most commonly used cultivars are Cabernet Sauvignon, Muscat Blanc, Chardonnay, White Riesling and Tinta Madeira. Two types of wine are prepared from grapes.

- ☆ Table wine: Alcohol content is less than 14 per cent.
- ☆ Dessert wine: Alcohol content of more than 14 per cent.
- ☆ The fruits should have 17-20 per cent sugar: acid ratio for better quality wine production.
- ☆ Wine grape producing areas are the USA, Europe, North and South Africa, South America and Australia.

5. Canning grapes: These are used to keep grapes and mixed with pieces of other fruits for mixed fruit cocktail. Commonly used cultivars are Pachadraksha, Anab-e-Shahi, Muscat (Panneer) and Bangalore Blue.

Origin

Origin is between Caspian and the Black Sea. (Armenia – a district near the Caspian Sea in Russia).

Soil and Climate

- Grapevine has a powerful root system that can be grown on a wide range of soils. Sandy loam soils (pH 6.5-7.5) with good drainage and fairly fertile with good organic matter are best suited. Moderately tolerant to salinity (up to 6 dSm^{-1}), however, the presence of excessive lime is harmful. Heavy soils, very shallow soils and ill drained alkaline soils are not suitable.
- A hot dry summer and cold winter climate is ideal. Bright sunny days and cool nights during the fruiting period contribute to better development of sugar and colour in berries. High night temperature (>25°C) during ripening hampers colour development in coloured grapes. High temperature also results in the formation of berries with thick skin. The optimum temperature range is 28 to 32°C.
- The vines remain dormant during winter in North India.
- Haryana has marginal climatic conditions because of the early onset of monsoon rains, which decrease heat accumulation, resulting in poor quality (less sweet and thick skinned) berries. Under such a climate, short-duration varieties with an early bearing habit suit well. Rains during ripening cause berries to crack and rot.
- In south India, grape vines remain evergreen and two crops are taken in a year (February-March and October). Showers or rain during flowering is very dangerous to grapes and reduces yield to a greater extent.

Cultivars

There are thousands of cultivars of grapes most popular ones are:

- **Coloured seeded grapes:** Bangalore Blue (*V. vinifera* x *V. labrusca*) is used for table, juice and wine purposes; Kishmish Charni and Black Champa for processing and hybridization purposes and Gulabi (Muscat) is used for table purposes.
- **Coloured Seedless:** Beauty seedless (blue, black) for early season; Krishmish Charni (brick red) for mid-season; Sharad seedless, Flame Seedless (Purple) for late season.
- **White seeded:** Anab-e-Shahi is a late maturing, seeded variety, it has been acclaimed as one of the most productive cultivars. It has attractive large bunches of berries with good shipping quality.
- **Dilkush** (clone of Anab-e-Shahi) and **Cheema Sahebi** (Selection from Pandari Sahebi) are late cultivars. Cheema Sahebi was selected by Dr G.S. Cheema. Its shipping quality is poor.

- **White seedless:** Perlette (Scolokertekhiralynoje-26 x Sultania Marble), developed by Dr HP Olmo. The most striking feature is the translucence.
- **Pusa Seedless:** Superior to Perlette (Mid-season), it resembles the Thompson Seedless in many characters.
- **Thompson Seedless:** It is a mid-season variety and was first grown by William Thompson. It is also called oval Kishmish. It is a table and raisin cultivar as well. Thompson Seedless and its clones are the major white seedless grapes grown in approximately 70 per cent of the area. Other seeded and seedless cultivars are Kishmish Chorni (Shared Seedless). Flame Seedless, Red Globe, Perlette *etc.* cover most of the rest area.
- Tas-A-Ganesh, Manik Chaman and Sonaka are clones of Thompson Seedless.
- Delight is the sister seedling of Perlette and early maturing.

Hybrids

- **Homorod:** Contario x Sultania.
- **Arkavati:** Black Champa x Thompson Seedless (late type).
- **Arka Chitra:** Angur Kalan x Anab-e-Shahi (tolerant to powdery mildew).
- **Arka Kanchan:** Anab-e-Shahi x Queen of Vineyard.
- **Arka Neelamani:** Black Champa x Thompson Seedless.
- **Shweta Seedless:** Anab-e-Shahi x Thompson Seedless.
- **Arka Majestic:** Angur Kalan x Black Champa.
- **Arka Krishna:** Black Champa x Thompson Seedless.
- **Arka Soma:** Anab-e-Shahi x Queen of Vineyards.
- **Arka Trishna:** Bangalore Blue x Convent Large Black.
- **Pusa Aditi:** Banqui Abyad x Perlette.
- **Pusa Trishar:** Hur x Bharat Early x Beauty seedless.
- **Pusa Swarnika:** Hur x Cardial.
- **Manjari Medika:** Pusa Navrang x Flame Seedless.

Propagation

All the grape cultivars are grown on rootstocks in Europe, USA, Australia and Russia due to soil-borne problems. In India, all the commercial cultivars are grown on their own roots as self-rooted cutting. The grape is best propagated by hardwood cutting. Propagation by hard wood stem cuttings of 20 to 25 cm in length and 0.75 to 1.25 cm diameter having 3 to 4 buds should be selected from well-mature canes of productive vines from December-January pruning. Finished cutting tied in bunches of 100-200 and buried in moist sand or saw dust for about a month for callusing. The callused cutting starts to swell in the nursery and gives a higher percentage of rooting. Plant the cuttings at a 15 cm x 30 cm (P x R) distance and after every two rows keep 60 cm space for cultural operations. The cuttings were treated with 500 ppm IBA for overnight and 2000 ppm IBA for a quick dip (10 sec) just before planting for better rooting. Rooting on single bud cuttings can be initiated by keeping them at 18 to 20°C for 2 to 3 weeks and chip budding or grafting can be done for vegetative propagation.

Rootstocks

Some important species used as rootstocks resistant to phylloxera and nematode are *Vitis riparia, Vitis rupestris, Vitis berlandieri, Vitis candicans, Vitis rufotomentosa* and *Vitis solaris*. The most commonly used rootstocks are as under:

- ☆ **1616:** Moderately resistant to phylloxera and highly resistant to nematode.
- ☆ **1613:** Used for Anab-e-Shahi or Thompson Seedless for nematode resistance and moderately resistant to phylloxera. Cutting roots readily and budding or grafting is also easy on these cuttings.
- ☆ **Dogridge:** Selection from *Vitis champini*, resistant to phylloxera, nematode and salts.
- ☆ **Salt Creek (Ramsey):** It belongs to *Vitis champini*, resistant to salt and nematode. Rooting is difficult, but budding and grafting is easy.
- ☆ **St. George:** Belongs to *V. rupestris*, vigorous root stock propagated from cuttings, grafts are successful and are highly resistant to phylloxera and tolerant to drought. This rootstock produces only staminate flowers.
- ☆ **Riparia Gloire:** Belongs to *V. riparia* and is resistant to phylloxera.
- ☆ **Temple:** Imparts resistance to scion against Pierce's disease, anthracnose and downy mildew.

Planting

One-year old rooted cuttings should be planted from January to mid-February before the vine starts sprouting. Dig a pit of 1 m^3 and fill it with a mixture of soil (upper half) and farmyard manure (50 kg) + 2 kg single super phosphate and add 30 ml chlorpyriphos to avoid termites. If the soil is clay, then add 50 kg of sand.

Planting Distance

- ☆ Head system: 6 feet x 8 feet (1.81 m x 2.4 m).
- ☆ Telephone: 8 feet x 10 feet (2.4 m x 3.0 m).
- ☆ Bower: 10 feet x 10 feet (3 m x 3 m).

Essential operations just after planting

As soon planting is over, apply irrigation. Protect the plants from sunlight by covering or shedding them. It is also advisable to give support to plants by fixing bamboo stakes near the base of each plant.

Manures and Fertilizers

In North India, grapevines become dormant and shed their leaves during winter. Application of manures and fertilizers is to be done soon after pruning at 45-60 cm away from the base of the trunk, depending on the age of the vine. Fertilizer should be added on the basis of leaf nutrient analysis and soil testing. Apply farmyard manure in the month of January, ½ urea and a full dose of single super phosphate should applied in the month of February just after the pruning, and the remaining half dose of urea along with potash should be applied in the month of April after fruit set.

In southern and western India, where winters are very mild, and hence, grapevines do not go under any dormancy or remain evergreen. These require pruning of vines twice in a year, *i.e.* April and October. Therefore, the application of manures and fertilizers should be added twice.

Table 12.1: Fertilizer Dose for Grape Vine Planted at 3 m x 3 m (North India)

Age of vine	*Dose kg/vine*			
	Farmyard manure	*Urea*	*Single super phosphate*	*Sulphate of potash*
2 years	30	0.250	1.00	0.500
3 years	45	0.375	1.50	0.600
4 years	60	0.500	2.00	0.800
5 years and more	75	0.675	2.00	1.000

Irrigation

- ☆ It is a shallow feeder and requires light and frequent irrigation for better yield. Irrigation frequency depends upon climate, soil type, age and condition of the orchard.
- ☆ 1st irrigation in February (after pruning and fertilizer application).
- ☆ 2nd and 3rd irrigation in March.
- ☆ 4th irrigation in April.
- ☆ It is very essential to irrigate the grapevine at an intervals of 5 to 7 days until the berry attains pea size and thereafter, irrigation is given at 10 days intervals till maturity.
- ☆ Grape requires less water during fruit bud formation and more water during berry growth. Reduced irrigations (little withhold) during ripening improve fruit quality. Drip irrigation is becoming more popular. Irrigation should never be ignored just after harvest.

Critical Irrigation Stages

- ☆ Grapevine should be irrigated after pruning and fertilizer application.
- ☆ Irrigation between fruit set to maturity is more critical.

- ☆ Water requirements under drip irrigation are as under:
- ☆ 1-40 days after summer pruning: 48,000-60,000 liters/day/ha.
- ☆ 41-100- days after summer pruning: 24,000-32,000 liters/day/ha.
- ☆ 101- days after summer pruning: 15,000-20,000 liters/day/ha.

Flowering and Fruiting

Excessive moisture and stress reduce berry setting. Apply irrigation before flowering, but avoid at full bloom stage. The flowers in grapes are borne in clusters as perfect, pistillate or staminate. In case of seedless varieties, berries develop by parthenocarpy, *i.e.* without fertilization. Seedless varieties of grape are Perlette, Beauty Seedless, Pusa Seedless, Delight, Thompson Seedless, in which fertilization occurs, but the embryo aborts after some time, this condition is known as stenospermocarpy.

Heat Unit Requirement

Heat unit is determined as a mean temperature (mean of maximum and minimum), which is calculated and deducts base temperature (10°C) from the mean temperature. The balance temperature is added from the bloom to the ripening of grapes. This is heat unit accumulation and expressed in terms of degree days for the ripening of a particular variety.

Heat Unit: $(T_{max} + T_{min})/2 - 10\ ^{\circ}C$.

Early ripening cultivars require 1600 to 2000-degree days, while late ripening cultivars require 3000 or more degree days.

Training

Training involves the removal of canes, shoots, leaves and other surplus vegetative parts of the vine to provide the desired framework of the grapevine. It enhances productivity by allowing agricultural operations to be easy.

- ☆ To distribute the proper amount of wood over the vine.
- ☆ To induce flowering and provide a suitable site for flowering and fruiting.
- ☆ Manage plants according to training systems head, kniffin, telephone and bower.

Training Methods for Grape Vine

i. Head System

Plants trained through this system develop in a bush shape. During early years, the vine requires support. The plant is allowed to grow to a height of 75-90 cm. After 3-4 years, the main stem becomes strong, and the vine takes the shape of bush *e.g.* Perlette and Beauty Seedless.

Fig 12.1: Head System

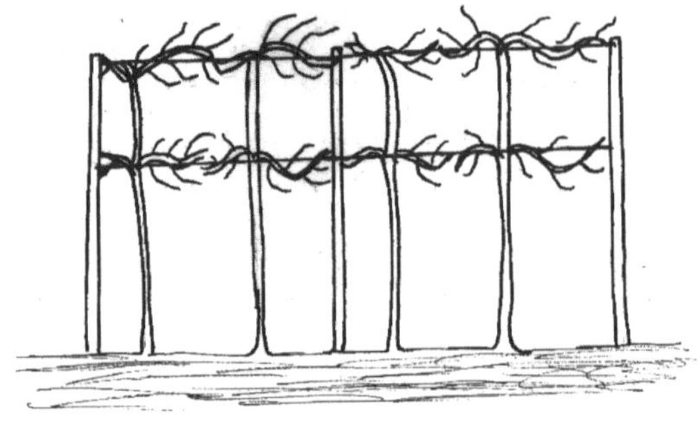

Fig 12.2: Kniffin System

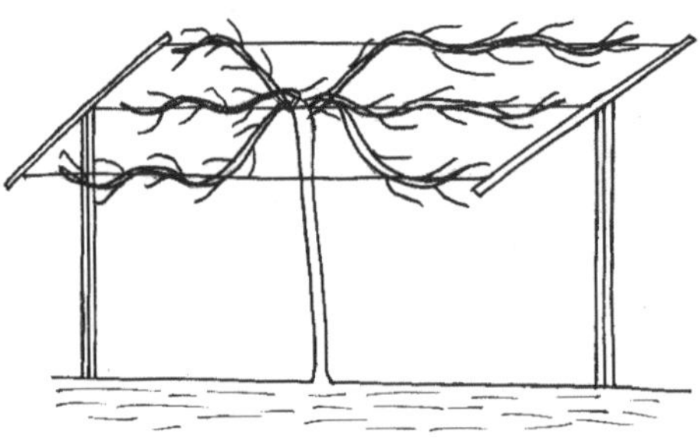

Fig 12.3: Telephone System

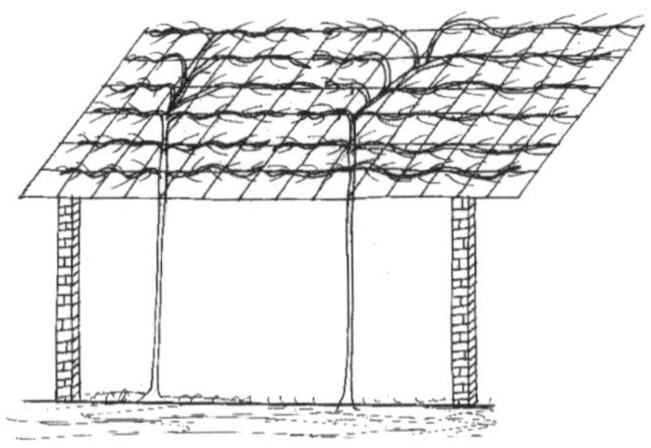

Fig 12.4: Bower System

ii. Kniffin System

In this system, two rows of wires are stretched at a height of 1.05 and 1.65 meters from ground level with the help of iron or concrete poll. The vines are allowed to grow on these wires.

iii. Telephone System/Overhead Trellis System

This system is known as the six canes system. The polls are erected at a distance of 3.6 to 4.8 m. At the terminal end of the poll, there is a 1.2 m long arm and wires on this terminal end arms are strung about 30-45 cm apart.

iv. Bower System

A network of wire is created and fixed 2.0 to 2.4 m above ground level at 45 cm apart on poles to develop a roof-like structure. This is also called an Arbour or Pergola. Planting distance 3 m x 3 m.

Precautions during Training

- ✰ Handle newly planted trees carefully.
- ✰ Select suitable and strong stakes only for sprouting.

- ☆ Avoid unwanted branches on the main trunk.
- ☆ Treat the wound with fungicide.
- ☆ Removal of unnecessary scaffold branches.

Pruning

Most technical and important operation for balance between growth and fruit production. In North India, vines are pruned in winter (December-January). Half of the canes are pruned to renew spur and the rest half for fruiting canes. One or two buds from a cordon (arm) are retained for renewal of spur and 3 to 8 buds are retained on fruiting canes. The buds on the fruiting cane can be retained on the basis of the fruiting capacity of the cane. The pruning should be done as under:

- ☆ Strong cane can be pruned more (longer).
- ☆ Weak cane pruned less (shorter).
- ☆ Fruiting cane pruned as renewal for fruiting in the next season.

Pruning Time

Mid-January is the best time for pruning when the vine is dormant, it should be completed before sprouting. The objectives of pruning are as follows:

1. Decrease the amount of old wood.
2. Make the size of the cane manageable.
3. Increase the number of bearing branches.
4. Make a balance between vegetative and reproductive growth.

Types of Pruning

Back pruning: Cutting back of canes to one to two bud levels for vegetative growth, generally practised in South India.

Fore pruning: Pruning of cane at 5 bud stage for table purposes and 3-4 bud stage for wine purposes.

Cane: This is the previous season shoot growth. Basal buds are generally vegetatively pruned, keeping 8-9 buds *e.g.* Thompson Seedless, Pusa Seedless and Kashmiri Charni. In cane pruned cultivars, two buds are left. However, in spur pruned cultivars, one bud is left.

Head system: Retained 8-10 canes.

Kniffin: Retained 16-20 canes.

Telephone: Retained 32-40 canes.

Bower system: Retained 40-50 canes.

Precautions: Canes should be 20 cm apart, well distributed canes should mainly located towards the centre of the vine. Cut the cane with sharp secateurs and apply Blitox at the cut.

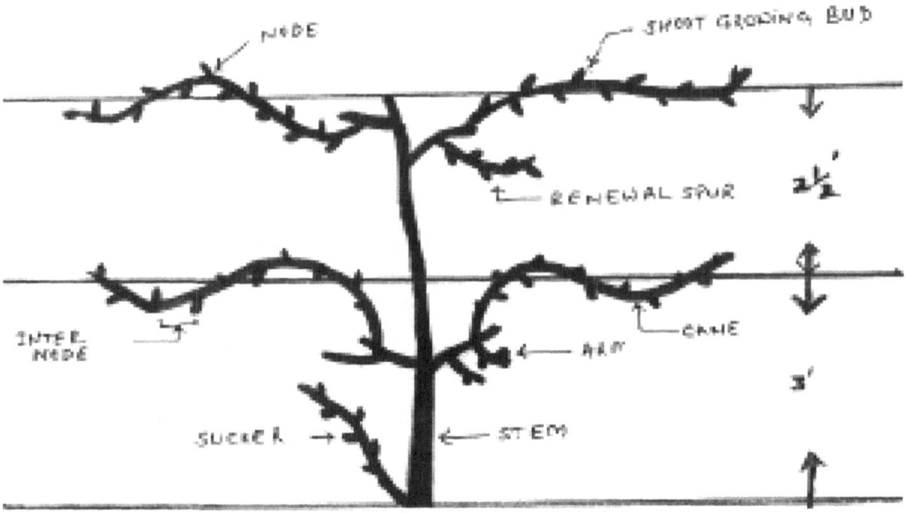

Fig. 12.5: Sketch of Grapevine

Rejuvenation and Top Working

- ☆ Heavy pruning or heading back, leaving the primary stubs with the objective of rejuvenate the plant. In North Indian conditions, this is done during the dormant period, *i.e.* December- January.
- ☆ After heading back, new sprouts can be grafted with new emerging cultivar(s) during the last week of February to improve the fruit quality *e.g.* Flame Seedless grafted on Perlette vine using tongue grafting. This training system has been adopted widely in India.
- ☆ Among different training systems on Thompson Seedless, Beauty Seedless and Himrod cultivars, the bower system of training gave 1.46 MT higher yield than the telephone, 1.70 MT higher yield than kniffin and 3.05 MT higher yield than head systems.

Terminology used in Pruning

1. **Shoot:** Current season growth of shoot with green bark.
2. **Spur:** Shoots left on plant after pruning *e.g.* In Perlette, Beauty Seedless fruits borne on 2-3 basal buds.
3. **Fruiting spur:** Cane or ripened shoot gives rise to the fruiting bunch.
4. **Foundation or Renewal spur:** The cane or stem left on the plant during pruning. These will produce cane or fruiting spur in the ensuing season. Vegetative growth will bear fruiting canes next year.
5. **Trunk:** The stem of a plant.
6. **Long spur:** The spur is long in size (2-30 cm) with more than 5 buds.
7. **Medium spur:** Retained 3-5 buds on the plant.

8. **Spur:** Retained 1-2 buds on the plant.

Weed Management

- The orchard should be weed-free and to reduce the seed bank, summer irrigation should be followed by tillage and repeat this operation.
- Mulching can also be done to reduce the weed intensity in the orchard, it prevents the germination of weeds. Mulching can be done with organic waste or plastic sheets.
- Chemical control: Apply glyphosate @ 2 kg a.i./ha or Diuron @ 2 kg/ha applied as post-emergence to control weeds.

Quality Improvement

- Application of GA_3 @ 20 ppm at full bloom stage and 40 ppm at pea size fruit stage increased berry size and yield of seedless cultivars. It avoids compactness in clusters, hence avoid rotting of clusters.
- Uniform colour development and early ripening in berries can be achieved by the application of Ethephon @ 500 ppm at the colour change stage (veraison stage).
- Pruning of bunch apex reduces the number of shot berries in the bunch and improves the quality of fruits in Beauty Seedless grapes.
- Spraying chlorocholine chloride (CCC) @ 500 ppm at the five leaf stage after back pruning increases fruitfulness.

Girdling

- Remove the 3-6 cm ring of the bark of trunk/branch/shoot. Girdling should not be done at the same point every year. The stage of girdling depends upon the purpose for which it is practised.
- To improve berry set and yield, girdling is done 10 days before bloom.
- To improve fruit quality and berry size girdling is done just after the fruit set.
- For better colour development, girdling is done at the veraison stage (Beauty Seedless and Kishmish Charni).

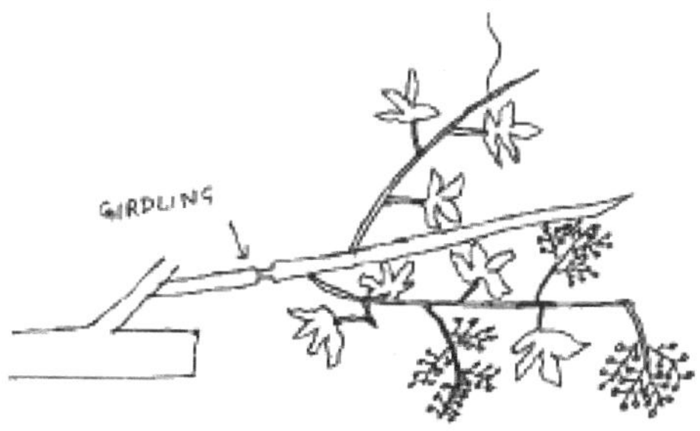

Fig. 12.6: Girdling in Grape

Thinning

On 40-50 canes there should be about 100 bunches. If more canes, more the chances of bunches misshapen. Small undesirable and over-crowded bunchs should be removed before fruit set. After the berry set, bunch thinning is not practised, but $1/4^{th}$ to $1/5^{th}$ of a bunch from the apex should be removed with scissors. Fruit thinning can be done easily with the application of GA_3 @ 50-75 ppm to increase berry size and quality.

Pinching

It is done to remove the succulent shoot tip (3-4 cm).

Topping

It is done by removing 10-12 cm or more portion of the shoot tip or terminal portion. Pinching and topping are not recommended in North India for quality improvement.

Harvesting

- Grape is a non-climacteric fruit, so fully mature fruit should be harvested. After harvesting, fruits are to be precooled/remove field heat for better shelf life. Grading and removal of rotten berries should be done after harvesting.
- The degree days requirement for harvesting of grapes is as under:

  1600-2000-degree days are required for early ripening cultivars.

  3000 or more -degree days are required for late-ripening cultivars.
- Spray 250 ppm ethrel 5 weeks after anthesis and 4 weeks after berry set for uniform colour development.
- Grapes start yielding from 2 to 3 years after planting and continue gives full yield up to 20 to 25 years. The average yield is about 25 to 30 MT/ha, but it can be achieved up to 60 to 75 MT/ha under better management practices.
- Grapes can be stored at 0 to 1°C for 7 to 12 weeks under a controlled atmosphere with 15 to 25 per cent CO_2.
- Reduced irrigation (withhold irrigation) 5-10 days before harvesting to improve quality and hastens ripening.
- The stage of berry start ripening is called veraison. Berries become soft.
- There may be a slight change in the colour of berries (veraison stage) and browning of the cluster stem.
- Formation of a waxy layer on the skin of berries and berries are easily detached from the stem on pulling.
- Seeds are loosely attached to pulp and seeds turn brown in colour.

- ☆ The taste of berries becomes sweet in most of the cases.
- ☆ The juice of the berries becomes thick and the refractometer reading (TSS) varies from 18 to 20 °B.
- ☆ **Raisin making:** Fumigation with SO_2 @ 10 g/kg fruit for 3 hours, followed by shade drying or alkali treatment followed by sulfuring and shade drying.
- ☆ **Wine making:** Select fruits according to colour – crush or make juice-adjust sugar (20-24°B) and acid concentration – add yeast- allow to ferment- allow sedimentation- remove sediment or yeast by siphoning.

Insects

- ☆ **Grapevine thrips (*Rhipiphorothrips cruentaus*):** It is a major insect of grapes and is also found on mango, jamun and guava. Its infestation increased during dry weather from April to June and August to November.

  The cultivars that have thick leaves and hairs on the lower side of the leaves are resistant to this insect. Spray 500 ml malathion (Cythion) 50 EC per acre or 150 ml fenvalerate (Fenval) 20 EC per acre in 500 liters of water.
- ☆ **Hairy caterpillar (*Euproctis spp.*):** They feed voraciously on leaves and also eat the fruit skin.

  Spray spinosad 45 SC @ 0.25 ml per litre of water.
- ☆ **Flea beetles:** They tend to avoid sunlight, so they remain hidden under the leaves during the day and feed on sprouting buds during the night and young ones feed on green stems and leaves.

  Spray malathion @ 1 ml per litre of water.
- ☆ **Chafer beetle:** Adult is reddish brown in colour and nocturnal in habit, so growers are unable to detect its presence and it feeds on leaves.

  Dusting with any insecticide during evening hours.

  Mealybugs: Their body is pinkish coloured and covered with white powdery wax. It feeds on leaves and fruits.

  Apply quinalphos @ 1.5 per cent.

Diseases

- ☆ **Anthracnose (*Gloeosporium ampelophagum*):** Dark brown spot around the midrib of the leaf. Later on, these spots also appeared on branches.

  The vine should be pruned as per recommendation and diseased branches should be removed and destroyed. Affected leaves, branches and plant parts should be burnt and disposed of. Spray Bavistin 0.2 per cent after pruning but before sprouting during dormancy. Repeat the spray in May, July and August. If it occurs even after the spray, repeat the spray up to mid September.

- ☆ **Downy mildew**: It is a serious fungal disease and infects all the green parts of the vine.

  Sprays Mancozeb @ 2.5 g/1 litre.

  Powdery mildew: Greyish white patches appear on the upper and lower surface of the leaves. It infects all the green tissues of the vine.

  Spray or dust sulphur or neem products.

Nematode

Root-knot nematode: Formation of the knot on the roots and root hairs and root starts rotting. The leaves become yellowish and stunted.

Apply Furadan 3 G @ 13 g/m^2 in 9 m^2 area per plant under the canopy of the plant. Mix it thoroughly and apply irrigation just after the application of Furadan. Grow garlic as a inter crop and decrease the dose of Furadan to half.

Physiological Disorders

i) Shot berries: Berries remain small, green, hard and don't ripen. It is more common in heavy bearer cultivars like Perlette, Delight and Beauty Seedless.

Causes: Non-judicious pruning, heavy bearing, malnutrition, poor pollination or fertilization, poor carbohydrate nutrition, boron deficiency, improper application of GA_3 *etc* cause the formation of shot berries.

Control

- ☆ Pre-bloom application of GA_3 @ 15 ppm.
- ☆ Avoid deficiencies of boron and Zinc.
- ☆ Berry thinning by the use of auxins.
- ☆ Dip the berries at the berry set stage in the solution of ethephon 25 ppm.

ii) Bud, flower and berry drop: It is also known as coulure or shelling. Excessive shedding of flower bud at about 8-10 days before full bloom causes loose and straggly branches. In cultivars like Thompsons Seedless, Beauty Seedless, Pusa Seedless and Gold, unopened buds and flower drops, as well as berry drops at the ripening or colour break stage.

Cause: Improper C: N ratio, hormonal imbalance, high ABA and low auxin, nutritional imbalance, climatic and management factors.

Control: Girdling to avoid over-cropping and excessive irrigation at flowering should be avoided.

Girdling on flowering bunches 10 days before full bloom.

Apply ethrel @ 500 ppm at ripening.

Spray NAA @ 100 ppm at 10 days before ripening.

Withhold or stop irrigation during flowering.

Fruit thinning by application of benzyl adenine @ 200 ppm and 4-CPA @ 20 ppm (CPA: chlorophenoxy acetic acid).

Prevent moisture stress and deficiency of Zn and boron.

Proper canopy management.

Interveinal chlorosis: Avoid deficiency of micronutrients like Mn, Fe and Zn.

Stalk necrosis: Calcium deficiency.

Bud killing: Excessive nitrogen.

iii) Hen and chicken: Presence of many small berries around a bold berry in a cluster. The bold berry is taken as indicative of a hen and the small berry is a chicken. Hence, it is called hen and chicken.

Causes: Impaired fertilization. Growth of many berries without embryo formation.

Zn and boron deficiency: In case of zinc deficiency, berry size is reduced, but the shape is usual. Whereas, in case of boron deficiency, the berry shape becomes spherical/oblate.

Control: Spray $ZnSO_4$ @ 0.5 per cent and boron @ 0.3 per cent as per schedule before flowering.

iv) Water berries: Berries with lack of firmness and become soft to feel during physical touch. Dull-coloured, watery berries which shrivel and dry by the time of harvest. This is a very serious problem in Thompson Seedless and Tas-A-Ganesh varieties.

Causes: Overload cropping and inadequate nourishment. Frequent application of nitrogenous fertilizer and frequent watering leads to excessive shoot vigour and continued shoot growth.

Control: Bunch and berry thinning, cluster clipping, limited watering and nitrogen application, use of potash and oil cakes, application of boric acid @ 0.2 per cent.

v) Pink berry: Development of a red streak on the peel of the few berries. Pink berries are developed in bunches during ripening before harvest. Turning from pink to a dull red colour renders the bunch unattractive. Pink pigment develops most commonly on Thompson Seedless berries due to more difference between day and night temperature (>20°C).

vi) Uneven ripening: Green coloured unripen berries on ripened branches.

Causes: Non-availability of reserve food due to inadequate leaf area. This is a problem in Bangalore Blue, Bangalore Purple, Beauty Seedless and Gulabi.

Control: Cluster thinning and stem girdling. Application of ethephon @ 250 ppm at colour break (veraison) stage.

Chapter 13

Guava (*Psidium guajava* L.)

Family: Myrtaceae; 2n = 22,33,44,66

Introduction

- It is a tropical crop but grown successfully under subtropical conditions.
- Guava is also called the apple of the tropics and poor man's apple.
- Guava is a self-pollinated as well as cross-pollinated crop. The ovary is inferior with many stamens.
- Fertilization in guava takes place before the opening of the flowers (Cleistogamy).
- The fruits of guava are available throughout the years as they bear fruits thrice a year (ambe bahar, mrig bahar and hasta bahar).
- It is a good and cheap source of vitamin C (about 220 mg/ 100g pulp) and is maintained during the processing of fruits. Fruits also contain Ca, P, Fe and thiamine.
- Fruits are good source of pectin, so they are the best option for the preparation of jelly.
- The edible portion of guava fruit is the thalamus and pericarp.
- About 50 per cent of the area under guava is in Uttar Pradesh. Allahabad area possesses a reputation for the best quality guava in India as well as in the world.
- Major guava producing states in India are Uttar Pradesh (23 %), Madhya Pradesh (15.8 %), Bihar, Andhra Pradesh, West Bengal and Chhattisgarh *etc.*

Important species

- *Psidium guajava* (common Guava) and *P. polycarpum* 2n=22.
- Anueploid no. 82 (Tetrasomic) 2n=33
- *P. guineense* and *P. cujavillus* 2n=44.
- *P. friedrichsthalianum* (Chinese guava) 2n=44,66
- *Psidium cattleianum (*Cattley guava*)* : 2n=66, 77.
- *Psidium guineense* (Brazilian guava) 2n=44.

Origin

Guava is native to tropical America and grown from Mexico to Peru. It was introduced in India in the 17th century.

Soil and Climate

- It can be grown in all types of soils, but loam and alluvial are best for its cultivation.
- Sandy loam soils with 5.5 to 7.5 pH are most suitable.
- For better yield, soil should be fertile, deep and well-drained without a hard pan.
- Soil should be free from nematode.
- This crop can be grown easily at an altitude of 1000-1500 m above sea level and the area receives about 500-1000 mm rainfall.
- Grown in tropical and subtropical climates.
- Trees need full sun exposure.
- Thrives well in a warm, humid condition but can withstand dry weather.
- Frost protection is required in colder areas, tree may die when the temperature goes below-3°C.

Propagation

- Guava is propagated both by seeds and vegetatively.
- Seedling trees produce fruits of variable size and quality, although such trees are generally long-lived.
- Vegetative propagation methods like inarching, wedge grafting and patch budding, air layering and stooling are generally preferred.
- For the vegetative propagation of guava, the rootstock is raised from the seeds extracted from the *desi* plant and L-49.
- Budding/grafting is done during July – September and February.

Rootstocks

- **Clonal rootstocks:** Mound layering is used for the propagation of guava rootstocks. In this method, first to establish a parent plant in the nursery.

Heading back the guava plant before spring or in the rainy season and remove the bark of the stool. Cover it with moist soil for rooting.

- ☆ **Seedling rootstocks:** Rootstocks are raised from the seeds extracted from healthy fruits. Sow the of seeds in March or August on a raised seed bed. Seeds could be stored for 104 days by presoaking in ferulic acid 10^{-3} M concentration. Soaking of seeds in water, followed by treatment with GA_3 @ 1000-3000 ppm improves the germination and growth of the seedlings. Seed germinates 12 days after sowing.
- ☆ Its seed has a hard coat and takes a longer time to germinate. Seeds are boiled for 5 minutes to reduce the time of germination.
- ☆ **L-49 (Sardar):** It produces good rootstock. Remove the seeds from the healthy fruits, wash and keep them in the shade for drying. If the seeds are sown just after extraction, then there are chances of high germination. The seeds are sown in the month of July-August in the raised nursery beds.
- ☆ **Chinese guava (*P. friedrichsthalianum*):** Dwarfing effect and resistant to wilt and nematode.
- ☆ **Pusa Srijan (Aneuploid No. 82):** Tetrasomic (2n+2), dwarfing effect on Allahabad Safeda shows resistance against guava wilt.
- ☆ ***P. pumilum*:** Dwarfing rootstock and increase the sugar content.

Varieties

- ☆ **Lucknow-49:** Also known as 'Sardar guava' developed by Cheema and Deshmukh at Pune by open progeny selection from Allahabad Safeda.
- ☆ **Allahabad Safeda:** A most popular variety of Uttar Pradesh.
- ☆ **Banarasi Surkha:** Fruits are of small size with pink flesh. However, it has a large number of hard seeds.
- ☆ **Lalit,** a guava variety having an attractive saffron yellow colour with occasional red blush and is medium sized with an average fruit weight. The flesh of the fruit is firm and pink in colour. It has a good blend of sugar and acid. It is suitable for both table and processing purposes. The pink colour in the beverage is made from its pulp, which remains stable for more than a year during storage.
- ☆ **Shweta** is a half-sib selection from Apple Colour with attractive fruits. Fruits are sub-globose with few soft seeds, high TSS (14 °Brix) and attractive pink flesh. It has good yield potential.
- ☆ Other new promising guava varieties are Shewta (CISH G4), Lalit (CISH G3), Lalima, Dhawal, Arka Mridula and Pant Prabhat.

Hybrids

Hisar Safeda: Allahabad Safeda x Seedless.

Hisar Surkha: Apple Colour x Banarsi Surkha.

Punjab Pink: Portugal x L-49 = F1 x Apple Colour.

Arka Amulya: Allahabad Safeda x Triploid.

Safed Jam: Allahabad Safeda x Kohir.

Kohir Safeda: Kohir x Allahabad Safeda.

Arka Kiran: Kamsari x Purple Local.

Hisar Safeda

Trees are tall and vigorous.

Fruits are round in shape.

Fruit skin is smooth, shining and yellowish white.

The flesh is white and firm.

Seeds are very few in number.

TSS is high, *i.e.* 12.5 – 13.0°Brix.

Hisar Surkha

Trees are spreading and are medium in size.

Fruits are round in shape.

Skin is light yellow in colour.

The flesh is pink in colour.

TSS is high *i.e.* 13.0 –13.5 °Brix.

Other hybrids: Arka Poorna (Punjab Local x Allahabad Safeda), Punjab Safeda (Shweta x 17-16) and Punjab Kiran (Apple Colour x 17-16).

Planting

Dig the pit of 1m^3 at the spacing of 6m x 6m in a square method. Fill these pits with 50 per cent top soil + 50 kg FYM + 3 kg SSP + 1.5 kg MOP + 50 ml chlorpyriphos. Keep these pits open for at least 1.5 to 2 months.

Ultra-High-Density Planting

- Plants spaced at 1 m x 2 m spacing accommodate 5000 plants per hectare.
- Plants are topped 2 months after planting in October for the emergence of new shoots below the cut ends.
- 50 per cent length of new shoots should be pruned in December-January for induction of more shoots, which enhance plant spread by May for flower bud differentiation.
- Production starts from the first year of planting (12.5 MT) and reaches up to 55 MT per hectare in 6 to 7 years.

☆ Lalit performs very well in ultra high-density planting (UHDP) systems.

Table 13.3: Comparison between Traditional and HDP Systems of Guava

Attributes	*Traditional systems*	*Meadow systems*
Bearing	After two years.	From first year.
Production	Overall production is low (12-20 MT ha^{-1}).	Increased overall production (30-50 MT ha^{-1}).
Management	Difficult to manage due to the large size of trees.	Easy to manage due to small size trees.
Labour requirement	More	Minimum
Production cost	Higher cost of production.	Reduced cost of production.
Harvesting	Difficult	Easy
Quality	Large canopy, poor sunlight penetration and poor quality.	A small canopy encourages air to circulate and sunlight to penetrate into the centre of the tree canopy, good air circulation minimizes disease and sunlight interception contributes to high fruit quality and colour.

Manures and Fertilizers

Table 13.1: Application of Manures and Fertilizers (kg/plant)

Age of Plant (years)	*Farmyard manure*	*Urea*	*Single super phosphate*	*Sulphate of potash*
1	10	0.200	0.200	0.100
2	20	0.400	0.400	0.200
3	30	0.600	0.600	0.300
4	40	0.800	0.800	0.400
5	50	1.000	1.000	0.500
6	60	1.200	1.125	0.500
7 and more	75	1.500	1.250	0.500

Irrigation

☆ New plantations require irrigation at weekly intervals during summer and fortnightly intervals during winter.

☆ Irrigation should be applied just after application of manures and fertilizers.

☆ Irrigation at the fruit development stage increases fruit size and yield.

☆ Timely irrigation also reduces the fruit drop.

Flowering and Fruiting

- ☆ Guava tree flowers throughout the year, but the peak flowering is observed in two seasons, the rainy season (April-May) and the winter season (August- September), and little flowering also occurs during November.
- ☆ Flowers are produced in leaf axils; the period of flowering varies from 25-45 days. Honey bees are working as pollinators.
- ☆ Under natural conditions, guava trees produce flowers and fruits twice a year in North India, but it is thrice, *i.e.* almost throughout the year in western and southern India. This pattern of flowering and fruiting is not desirable for commercial cultivation.

Crop Regulation

Table 13.2: Flowering Seasons in Guava

Bahar	*Flowering*	*Fruiting and ripening*	*Fruit quality*
Ambe bahar	April	July-August	Poor in quality and taste.
Mrig bahar	August-September	November-December	Fruits are of excellent quality.
Hasth bahar	November	March	Quality is good, but yield is low and fetches a good price.

Ambe Bahar (Rainy season crop): Insipid, less sweet, low keeping quality, infested with fruit flies.

How to avoid Ambe Bahar

- ☆ Withhold irrigation from mid February to mid May.
- ☆ Deblossoming: Manually or by the spray of 300 ppm NAA at full bloom in May (for flower drop). Dissolve NAA in a little amount of alcohol or spirit and then make the required volume with water.
- ☆ Spray 10-15 per cent urea in May (for flower drop).
- ☆ Pruning terminal portion of shoot up to 20-50 cm in April-May.
- ☆ Root exposure and root pruning are opted to avoid rainy season crops.

Rejuvenation of Old and Unproductive Guava Orchards

- ☆ Orchards lose their productivity beyond 25 years due to biotic stresses.
- ☆ Rejuvenation involves heading back of trees to a level of 1.0 to 1.5 m from ground level.
- ☆ Emerging shoots are allowed to grow (40 to 50 cm) for 4 to 5 months after heading back.
- ☆ Shoots are further pruned to 50 per cent of their length to allow the emergence of multiple shoots; these shoots increase fruit bearing surface.

Develop a healthy and productive canopy.

- ☆ Profuse fruiting and yield increased to about 70 to 90 per cent over non-rejuvenated trees from the first year of rejuvenation.
- ☆ Additional income from pruned wood and intercropping with short-duration vegetables and ornamental crops are the other advantages.
- ☆ Fruit size, yield and quality increased due to better light penetration into the canopy.

Insects

Fruitfly

- ☆ Ovipositional damage in the form of minute depressions may be seen from the outside. Fruit becomes soft at the site of infestation. The affected fruits rot and drop-down prematurely.
- ☆ To check the carry-over of the pest, collect and destroy fallen and infested fruits along with fruit fly maggots. Tillage of tree basins to check the pest population as pupae and hibernating larvae can be destroyed by natural enemies. Application of malathion 50 EC @ 1 ml per litre of water.

Mealybug

- ☆ Yellowing, withering, drying and shedding of leaves, drying of young shoots and dropping of affected fruits. Sooty mould develops on the honeydew like substance excreted by these insects, and it hinders the photosynthetic activity of the plant.
- ☆ The affected leaves and young shoots may be pruned and destroyed along with the pest during early fruiting or non-fruiting season. It helps in bringing down the initial pest population and prevents further spread. Overlapped and overcrowded branches may also be pruned to check the spread of these pests in the orchards. In case of heavy infestation, 0.5 per cent quinolphos in the early fruiting season or non- fruiting season is suggested. The spray may be repeated at fortnightly intervals depending on the severity of the attack.

Diseases

Wilt

- ☆ It is a fungal disease caused by *Fusarium* spp. The disease can be kept under check by proper sanitation in the orchard. It is most common in alkali soils.

  The wilted tree should be uprooted, burnt and trench should be dug around the tree trunk. Add organic manure, oil cakes and lime to check for disease. Drench with Bavistin @1 g/litre water.

Anthracnose

- Fruit and leaf infections are generally seen in rainy season crops. Pin head spots are seen on unripe fruits, which gradually enlarge. Spots are dark brown in colour, sunken, circular and have minute black stromata in the centre of the lesion, which produce creamy spore masses in moist weather.
- Spray Bordeaux mixture (3:3:50) just after initiation of disease. Spray copper oxychloride (0.3 %) just after the initiation of disease.

Chapter 14

Litchi [(*Litchi chinensis* (Gaertn) Sonn.]

Family: Sapindaceae; 2n = 28, 30

Introduction

- In India, litchi reached through Burma in the 17th century and was first introduced in Bengal.
- Litchi is an important subtropical, juicy fruit. It comes into the market in the month of May and early June when very few other fruits are available. Hence, it fetches good prices in the market.
- It is a single seeded fruit and the seed loses its viability in a very short period if exposed to natural conditions.
- Development of red colour on fruit skin is due to anthocyanin.
- In world production, India is 2nd highest producer of litchi after China.
- Important litchi growing states in India are Bihar (40 %), West Bengal (16 %), Jharkhand, Assam, Chhattisgarh and Uttarakhand.
- It is being cultivated in many countries of the world such as India, West Indies, Myanmar, South Africa, Australia, Thailand, Hawaii, Hong Kong and Mauritius.

Origin

- It is native to southern China.

Soil and Climate

- Litchi needs fairly deep, well-drained loam soil, fertile and rich in organic matter for its better production. However, it can survive on a wide range of soils.

- Soil pH 5.5-7.0 is suitable. In high pH and saline soil, it is difficult to establish and flourish.
- It is a sub-tropical fruit, so it flourishes best in a moist atmosphere in the subtropical climatic zone.
- Frost and dry hot winds are not favourable for the successful cultivation of litchi.
- Rainfall during flowering reduces fruit set.
- The optimum temperature is 30°C, while a higher temperature causes fruit cracking.

Cultivars

- **Dehradun:** It is an early bearing variety that ripened in the second week of June and grown in Haryana, Punjab and Uttrakhand. It bears regularly and heavily.
- **Calcuttia:** It can be grown successfully in hot and dry areas. Fruits are large-sized, attractive and mature during the third week of June. Fruits are less susceptible to cracking and sun-burn.
- **Seedless Late:** Seeds are shrivelled and flesh proportion in the fruit is more. Its fruit matures in the third week of June and are more prone to splitting than Calcuttia.
- **Rose Scented:** Fruit of this variety have a distinct rose aroma. Fruits have attractive pink skin. Fruits are mostly heart-shaped, the pulp is greyish white, soft and very sweet.
- **Saharanpur Selection:** This is an early maturing variety that ripens in the first week of June and yields heavily. The fruits are large, heart-shaped and deep orange to pink in colour.
- **Muzaffarpur:** The fruits are deep orange to pink in colour and are less prone to splitting. The fruits mature in the middle of June in northern India.
- **Shahi:** It is an important cultivar of litchi, matured in the 3rd week of May. It was GI (geographical indication) certified from Muzaffarpur, Bihar, in 2018. It is famous for its aroma, sweetness and pulpiness. It is suitable for canning purposes.
- **Sabour Lichi 1:** Fruits mature in the first week of June and are free from cracking.

Propagation

- Commercially propagated by air layering in February- March and July-October. Air layering is called Gooti in India and Marcottage in China.
- Mid-October is the best time for air layering. For early rooting, the upper cut of layering is treated with 500 ppm IBA.

- ☆ Stem cutting and semi-hard wood cuttings treated with IBA @ 3000 ppm for early rooting and stooling are practised.
- ☆ Seed propagation for raising rootstock: Fresh seeds are sown just after extraction, as the seeds of litchi have short viability for 4-5 days only.
- ☆ Seedlings of 12 to 18 months old serve the purpose of rootstock.
- ☆ Seeded trees come into bearing in 9 to 12 years and normally don't breed true to type.
- ☆ *Litchi philippinensis*, wild grown in Philippines is used as rootstock.

Planting

- ☆ A pit of 1 m^3 is to be filled with 30-40 kg farmyard manure + 2 kg bone meal + 300 g muriate of potash along with 50 per cent top soils about 2 to 3 months before planting.
- ☆ Litchi is planted in square system of planting at 8 m x 8 m (budded) or 10 m x 10 m (Seedling) apart.
- ☆ Planting of litchi should be done during the rainy season or after the rainy season. Litchi can be planted during the spring season also but planting during September- October is more successful.
- ☆ Budded plants start bearing in 5 to 6 years and seedling starts bearing 9 to 12 years after planting.

Manures and Fertilizers

It is a heavy feeder crop. Deficiency of major nutrients (N, P and K) resulted in stunted growth and may hinder floral initiation. So for better production, nutrients should be applied timely in the required dose as given below:

Table 14.1: Application of Manures and Fertilizers (kg/plant)

Age of plant (years)	*Farmyard manure*	*Urea*	*Single super phosphate*	*Muriate of potash*
1-3	10-20	0.2-0.5	0.2-0.5	0.06-0.150
4-6	25-40	0.5-1.0	0.75-1.25	0.2-0.3
7-10	40-50	1.0-1.5	1.25-2.0	0.3-0.5
>10	60	1.75	2.25	0.6

The basal dose of farm yard manure, single super phosphate, muriate of potash should be applied in the month of December. Apply urea in two split doses, the first half in mid-February and the second half after the fruit set in mid-April.

Irrigation

- ☆ Young plants, as well as bearing plants need regular irrigation.
- ☆ The fruit development phase requires regular irrigation, *i.e.* from March to May.

- Irrigation should be given after the fruit set and it is not required before the fruit set.
- A dry period or lack of irrigation results in fruit drop and splitting of fruits.

Protection from Frost and Hot Winds

- Protection of young plants from frost and hot winds: Grow *dhaincha* in the basin around the plant in February- March for protection from hot winds. The top of the *dhaincha* is tied in winter to protect against frost. Open space between *dhaincha* stems can be plugged with grass or straw. Roots of *dhaincha* are pruned to avoid competition with litchi roots.
- Planting of windbreak.

Harvesting

- It is a non-climacteric fruit, so it should be harvested only after attaining full maturity.
- The fruits are harvested during May and June along with branches and a few leaves for longer storage life or to avoid the rotting of fruits from the peduncle attachment sites.
- On a computation basis, its maturity can be judged by attaining 55-60 days from flowering.
- Fruits have a short shelf life, so the fruits can be stored up to 2-3 days under ambient temperature.
- Fruits can be stored for 3-4 weeks at 5-7°C.
- Fruit colour can be improved by dipping in 250 ppm ethrel.

Insects

Mites: Adult and nymphs suck leaves sap and cause curling of leaves.

Spray 0.1 per cent dimethoate (Rogor) at fruiting or malathion @ 0.1 per cent at fruit maturity.

Leaf roller: The larvae feed on leaves and roll the leaves during feeding.

Remove the infested leaves during the initial phase of infestation. Spray 0.1 per cent malathion.

Bark-eating caterpillar (*Indarbela quadrinotata*): Caterpillar make holes in the stem and feeds tissues inside the hole during the day while feeds outside the hole during the night. It is mostly found in unmanaged and old orchards.

Apply 0.1 per cent monocrotophos or Kerosene in holes.

Diseases

Anthracnose: Circular to irregular shaped brownish spots formed in the pericarp with well-defined margins.

Spray copper oxychloride @ 0.3 per cent.

Fruit rot: It starts from the injured portion of the fruit. Initially, the affected area gets depressed and later on penetrates into the pulp.

Spray 0.1 per cent Bavistin.

Physiological Disorder

Fruit Cracking

It is one of the most common problems of litchi. There are so many reasons for fruit cracking, but high temperature, low humidity and low soil moisture conditions during fruit development promote this disorder. Fruit crack happens when the internal pressure increases as a result of rapid aril growth after irrigation or rains, but fruit skin becomes hard.

Control

- ☆ Low-headed plants with dense foliage can withstand more against hot and desiccating winds than tall trees. Fruit splitting is least in dense foliage.
- ☆ Apply frequent and adequate irrigation. Mulching of plants and grow windbreak plants to avoid damage by high wind velocity.
- ☆ Spray NAA or 2,4,5-T @ 35-100 ppm. Spray Borax @ 0.8 per cent or $ZnSO_4$ @ 1.5 per cent to reduce fruit cracking.

Chapter 15

Papaya (*Carica papaya* L.)

Family: Caricaceae; 2n = 18

Introduction

- ☆ In India, it was introduced by the Portuguese in 1611.
- ☆ Papaya is a tropical fruit, ideal for growing in a kitchen garden as well as a filler crop.
- ☆ It is a rich source of vitamin A (2500 IU) and contains valuable proteolytic enzyme papain, which helps in the digestion of protein. It is rich in food and is also used for manufacturing chewing gum, cosmetics, meat tenderizing leather, silk and wool industry. Unripe fruits are consumed as vegetables.
- ☆ Papaya has been known as the '*Wonder fruit of the tropics*', *Melon tree* and *Papaw*.
- ☆ Papaya is a dioecious plant and sometimes trioecious means separate plant bearing either male, female or bisexual flowers.
- ☆ The yellow pigment in papaya is due to caricaxanthin.
- ☆ Papaya is cultivated in India in 1.5 lakh ha and its production is 53.4 lakh MT. This crop has the highest productivity after bananas.
- ☆ Papaya seeds are enclosed with gelatinous sarcotesta.
- ☆ India ranks 2nd in the area and 1st in production (45 %) in the world. The productivity of papaya in India is at 9th position, *i.e.* 43.4 MT/ha in the world.

- Papaya growing states in India are Andhra Pradesh (28.2 %), Gujarat (21 %), Karnataka, Madhya Pradesh, Maharashtra, West Bengal, Chhattisgarh, Madhya Pradesh, Telangana, Tamil Nadu, Assam, and Bihar.
- It is commercially grown in so many countries of the world. It is mainly grown in India, Brazil, Indonesia, Nigeria, Mexico, Taiwan, Australia, Peru, Hawaii, Florida, Puerto Rico, California, Texas, Pakistan, South Africa and Bangladesh.

Origin

- Papaya is native to tropical America (Mexico).

Soil and Climate

- Fertile, well-drained sandy loam soils are best suited for the cultivation of papaya.
- The pH of the soil should be within 6 to 7.
- Calcareous and stony soils with poor organic matter are not suitable.
- It is a tropical fruit that requires a warm and humid climate but also grows well in the mild subtropical regions.
- Among different climatic factors, temperature is the most important factor in determining the success of papaya cultivation. Temperature below 10°C affects adversely on growth and fruit development. It also delays maturity and ripening.
- The optimum range of temperature for papaya is 22-26°C. High temperature tends to induce more maleness.
- It is very sensitive to frost, strong winds, scorching heat and water stagnation. Frost is the most limiting factor in North India.

Cultivars

- **Pusa Delicious:** Gynodioecious, fruit size is large. The average fruit weight is 1 to 1.2 kg.
- **Pusa Dwarf:** It is a dioecious variety. Plants are dwarf and suitable for high-density planting. The plant starts bearing 40 cm above the ground level. The average fruit weight is 1.2 kg.
- **Pusa Giant:** Dioecious, plant start bearing at 1 m height. The fruit size is 2.5 to 3.5 kg.
- **Pusa Nanha:** Dioecious dwarf variety developed at Regional Station, IARI, Pusa, Bihar. It is a mutant, developed through gamma irradiation from the seeds of a local variety (mutant variety).
- **Washington:** Dioecious, plants are fairly dwarf and fruit size is medium.
- **Honey Dew:** Dioecious, semi-tall variety, also known as Madhu Bindu.
- **Croog Honey Dew:** It is gynodioecious and a selection from Honey Dew. The plant is tall and bears large fruits.

- **Pusa Majesty:** Gynodioecious, tolerant to viral disease and root-knot nematode.
- **Surya:** Gynodioecious, developed at IIHR, Bangalore (Sunrise Solo x Pink Flesh Sweet).
- **Sunrise Solo:** Gynodioecious, the solo name of this variety was kept on the basis of its size, as one man can easily consume one fruit. It produces no male plants.
- **Red Lady:** This is a Taiwan hybrid and has been adopted on a large scale.
- **Pant 1, Pant 2 and Pant 3** were developed from Pant Nagar.
- **Co-1 to Co-7:** Released from Tamil Nadu Agricultural University, Coimbatore (all dioecious except CO 3).
- **Gujarat Junagadh Papaya 1 (GJP 1):** This variety has less incidence of papaya leaf spot.
- **Phule Vijaya (Sel. GKPS-2-7):** Resistant to papaya ring spor virus disease.

Propagation

- Papaya is usually propagated by seeds.
- The seed rate is 250-300 g/ha for gynodioecious and 400-500 g/ha for dioecious varieties.
- Sowing 120 g seeds in a nursery area of 40 m^2 is sufficient for transplantation in one acre. Add one quintal FYM fin this area before sowing seeds. Seed should be sown at 2 cm depth and 15 cm spacing in April.
- Under air tight conditions, the seed can be stored for 45 days.
- Treat the seed with Captan @ 3 g/kg seeds. Drenching with 0.2 per cent Captan in a nursery bed or polybags to protect against damping off.
- Seed treatment with thiourea @ 100-200 ppm or GA_3 @ 200 ppm improves seed germination.
- Plants are ready for transplanting in May –June.

Tissue culture: Plants can also be propagated by tissue culture. In this method, a large number of true-to-type plants can be multiplied from a limited planting material in a very short period.

Planting

- Planting of this crop can be done thrice in a year, *i.e.* spring season (February -March), monsoon season (June-July) and autumn season (October-November).
- Spring season planting is followed in the areas of mild climatic conditions throughout the year. Monsoon season planting is followed in frost prone areas, especially in northern India, and autumn season planting is done

to avoid the crop from water stagnation (due to high rainfall) and virus problems.

- The pit size is 50 cm x 50 cm x 50 cm filled with 30-40 kg FYM with 25 ml chlorpyriphos with top 50 per cent soil in June. A spacing of 1.8 m x 1.8 m is normally followed for planting in most of the cultivars.
- Pusa Nanha is suitable for high-density planting (spacing 1.2 m x 1.2 m).
- Due to sex variation, more than one per pit is required.
- Irrigate the plant just after planting and provide staking to the plant.

Removal of unwanted male plants: Keep 10 per cent of male plants in dioecious cultivars for pollination.

Frost protection: Young plants should be covered with rice straw/sarkanda/ white polythene bag; secondly, irrigate and smoke the field to protect against frost.

Manures and Fertilizers

- It is a heavy feeder crop.
- Make a mixture of fertilizer *viz.*, urea, single super phosphate (SSP) and potassium sulphate in a ratio of 1: 4: 1.
- Apply ½ kg mixture in the month of February and the remaining half in August, along with 20 kg FYM about 35-40 cm away from the main stem.

Irrigation

Irrigation frequency depends on the types of soil, climate and plant growth stage.

- During the winter season, irrigate the plants once in 2 weeks while once in a week during summer.
- Avoid water stagnation, particularly near the stem. Papaya cannot tolerate wet feet.
- Heavy irrigation and water in direct contact with the stem should be avoided. Otherwise, the plant may suffer from root rot and collar rot.

Sex Distribution

- Environmental factors affect the formation of different types of flowers, *i.e.,* male or female. Lower temperatures produce the female flowers on the male plant.
- The fertile and hermaphrodite nature of flowers also changes with environmental conditions, *e.g.* plants produce male flowers during summer and female flowers during winter.
- The variation in day and night temperatures also produces more female flowers as compared to normal flowers.

- ☆ However, in Coorg Honey Dew, high temperatures and long days promote female flowers.
- ☆ Application of GA_3 at 25 ppm and ethrel @ 100 ppm on 100-125 days old seedlings produce more female flowers.
- ☆ Defoliation has been reported to induce maleness.
- ☆ Papaya is polygamous; there are 3 important types of flowers reported by Frankel and Galun, 1977.
  1. **Staminate:** Long, narrow tubular flowers on long peduncles.
  2. **Pistilate:** The flower forms close to the trunk or with a very small peduncle, which is large in size, yellow in colour, borne singly or in groups of three in the leaf axils.
  3. **Hermaphrodite:** Creamy white, big-sized flowers, without peduncles close to the trunk in leaf axils.

Harvesting

- ☆ Papaya is a climacteric fruit and the pattern of fruit growth is double sigmoid.
- ☆ Fruit is a fleshy berry.
- ☆ Papaya took 145 to 165 days to attain maturity from the date of flowering.
- ☆ Bearing starts in 8-10 months in South India and 1½ years in North India.
- ☆ At maturity, latex of the fruit becomes watery instead of milky. The fruit colour also changes from green to greenish-yellow.

Ripening and Storage

- ☆ For even and complete ripening, dip the physiologically matured fruits in 500 ppm ethephon for 10 minutes and wrap them in a newspaper after removing the surface water droplets.
- ☆ Ripening of papaya in calcium carbide (CaC_2) is dangerous as it may cause cancer.
- ☆ The optimum temperature for ripening and storage of papaya is 20°C.
- ☆ Grading of the fruit is done on the basis of size, weight and colour.
- ☆ Fruits can be stored for 1 to 3 weeks at a temperature of 10 to 13°C and 85 to 90 per cent relative humidity.

Insects

Aphids: It is sucking pests and suck the sap from the under surface of the leaf. In severe conditions, leaves become curled and heavily infested leaves fall down.

Spray 0.1 per cent malathion.

White fly: The female lays eggs on the lower side of leaves. Nymphs and adults both suck the leaf sap. The leaves become yellowish and drop off.

Spray 0.1 per cent malathion.

Red spider mite: Mites extract cell content from the leaf and turn it into white to yellow due to reduced chlorophyll content.

Spray 0.1 per cent malathion.

Diseases

Collar rot and wilt: The pathogen causes a soft, watery lesion near the collar region of the stem. In severe damage, plant falls down.

Avoid water stagnation, destroy affected plants and seed treatment.

Anthracnose: Formation of a pink-coloured spot on the fruit surface. Later on, the size of spot increases and becomes sunken. The sunken area turns black in colour.

Spray 0.2 per cent Captan or Dithane M-45.

Damping off: This disease occurs in the nursery. Formation of water-soaked spots at the collar region or close to the ground and seedlings may die during the severity of the disease.

Seed treatment with Captan @ 3 g/kg seeds and drenching with Captan @ 0.2 per cent.

Viral Diseases

Leaf curl virus: The leaf margin rolls inward and downward. Leaves become leathery, with thick veins, and stunted plant growth. No fruit and flower formation occurs. Whiteflies act as vectors to spread leaf curls.

Control the population of white flies and destroy the affected plants.

Nematode

Root-knot nematode: The plant affected with root-knot nematode reduces its yield.

Apply neem cake @ 1 kg/plant.

Chapter 16

Apple (*Malus domestica* Borkh): King of Temperate Fruits

Family: Rosaceae, 2n = 34

Introduction

- ☆ Apple is the premier table fruit. India ranks 2nd in area after China and 7th in production in the world.
- ☆ 'An apple a day keeps the doctor away'. This is the old proverb which signifies the nutritive importance of apples in the human diet.
- ☆ Apple is a rich source of easily assimilable carbohydrates (13.4 %), and it is fairly rich in calcium, phosphorus and potassium.
- ☆ The low productivity is due to poor fruit setting on largely grown delicious varieties. The reasons for low productivity are:

  Low pollinizer plant ratio.

  Prevailing low temperature at the time of flowering.

  Fruit setting coupled with irregular bearing tendency.

  Low yield potential.

- ☆ Botanically, an apple is a false fruit (pome) with a fleshy edible thalamus.
- ☆ Red Delicious is the mother of all Delicious groups of cultivars.
- ☆ Himachal Pradesh is called the 'Apple Bowl of India.
- ☆ Major apple producing states in India are Jammu and Kashmir (78 %), Himachal Pradesh (19.2 %), Arunachal Pradesh, Kerala and Nagaland.

- China is the largest producer of apples in the world, it produces about 45.5 per cent of the total world production.

Origin

- *Malus domestica* is a cultivated apple and it is believed to be originated in south-western Asia in the Caucasus region in Turkistan.

Soil and Climate

- Loamy soils rich in organic matter with a pH of 5.5 to 6.5 having proper drainage and good aeration are most suitable.
- Soil should be free from hard substrata and water-logged conditions.
- Apple is essentially a temperate crop and it is grown in temperate regions of the world. It can also be grown under a subtropical zone at an altitude of 1600 – 2500 m above mean sea level (msl).
- Low temperature during the bud rest favours better crop production.
- Most of the apple cultivars require uninterrupted 1000- 1500 hours of chilling below 7 °C.
- The average summer temperature should be around 21-24°C during the active growth period.
- Frost-free spring and adequate sunshine during summer without much fluctuations in temperature are most suitable.
- During flowering, bee activities are hampered by low temperatures as well as rains, cloudy weather, high and dry winds.
- The optimum temperature for pollen germination and fruit set is 21.1 °C to 26.7 °C.
- Excessive rains and fog at the time of fruit maturity invite diseases.

Cultivars

- **Commercial varieties:** Red Delicious, Starking Delicious (Royal Delicious), Richard, Mollies Delicious, Vance Delicious, Top Red, Granny Smith.
- **Colour strains:** Top Red, Vance Delicious, Hardeman, Red Gold.
- **Spur type:** Red Spur, Red Chief, Starkrimson, Orange Spur II, Gold Spur, Silver Spur, Well Spur, Scarlet Spur.
- **Low chilling:** Vered, Tropical Beauty, Michal, Schlomit, Neomi, Anna, Hariman, Dorsett Golden, Gordon.
- **Scab resistant:** Sir Prize, Priscilla, Prima, Liberty, Redfree, Coop 12, Freedom, Macfree, Jonafree.
- **Early season varieties:** Mollie's Delicious, Michal, Maayan, Shlomit, Anna, Chaubattia Anupam.

- **Mid-Season Varieties:** Red Delicious, Vance Delicious, Richared, Royal Delicious, Starkrimson, Red Chief, Well Spur, Orange Spur, Red Spur, Gala Selection, Hardeman, Bright-N-Early, Golden Spur, Silver Spur.
- **Late season varieties:** Lal Ambri, Firdous, Top Red, Red Delicious, Sunehari, Golden Delicious, Ambri, Fuji, Red Gold, Yellow Newton.
- **Pollinizing varieties:** Tydeman's Early, Red Gold, Golden Delicious, McIntosh, Winter Banana, Golden Spur.

Hybrids

- Lal Ambri: Red Delicious x Ambri.
- Sunehari: Ambri x Golden Delicious.
- Chaubattia Princes: Early Shanburry x Red Delicious.
- Chaubattia Anupam: Early Shanburry x Red Delicious.
- Ambred: Red Delicious x Ambri 157.
- Ambrich: Richared x Ambri 15.
- Ambroyal: Starking Delicious x Ambri 84.
- Fuji: Red Delicious x Ralls Janet.
- Ambstarking: Starking Delicious x Ambri 81.
- Pusa Gold: Golden delicious x Tydeman's.
- Pusa Amartara Pride: Royal Delicious x Prima.

Area Specific Cultivars

- **Himachal Pradesh:** King of Pippins, McIntosh, Golden Delicious, Red Gold, Starking Delicious, Yellow Newton and Granny Smith.
- **Jammu and Kashmir:** Irish Peach, Cox Orange Pippin, Kerry Pippin, Ambri, Lal Cider, Golden Delicious, Lal Ambri, Red Delicious, Sunehri and Razakwar.
- **Uttar Pradesh:** Fenny, Cortland, Early Shanburry, Golden Delicious, McIntosh, Red Delicious and Buckingham.
- Parlins Beauty and Tropical Beauty have been judged as the best performers in the hills of tropical regions where warm winter conditions exist. They require less chilling for bud break and flowering.

Rootstocks

- The East Malling Research Station in England developed Malling stocks (M). John Innes Horticultural Institute, Norwich, UK, Merton and East Malling Research Station, Kent, England, jointly developed the Malling Merton (MM) series of rootstocks, clonally propagated either by cutting or stooling. The rootstocks are easily propagated by hardwood cutting, treated with a growth regulator IBA @ 2500 ppm.

- **M 7, MM-106:** Tolerant to low temperatures below freezing (-40°C to –35°C). MM series of rootstocks are resistant to apple woolly aphids.
- **Seedling rootstocks:** Apple seeds require stratification at 4 to 7°C for 60-90 days. Apples are propagated on Crab Apple (*Malus sylvestris*) seedlings or self-pollinating varieties Golden Delicious, McIntosh, Yellow Newton and Northern Spy.

Clonal rootstocks

- Very Dwarfing: M27.
- Dwarfing: M9.
- Semi dwarfing: M26, M7, MM106.
- Semi vigorous: MM104, MM109, MM111.
- Vigorous: M12, M16, M25.

Propagation

- Tongue grafting during February – March.
- 'T' Budding in August.
- Chip budding in August.
- For apple seedlings that fail to succeed with tongue grafting during February-March, shield budding should be done in the month of June or chip budding during mid-March.
- Propagation by stooling, cutting and seedlings is used for the production of rootstocks.

Planting

- Planted in the month of January-February by a square/hexagonal/contour system of planting at a distance of 1.5 m to 6.0 m for vegetatively propagated plants and 7.5 m spacing for seedlings. Spacing depends upon various factors such as cultivars, rootstock used, tree size, soil type, soil fertility, irrigation facility, climatic conditions, planting system *etc.*
- Most of the varieties are self-unfruitful, so pollinizer varieties should be planted in between. Usually, one in six or one in nine plants is used as a pollinizer in self-unfruitful varieties.
- Low density: Less than 250 plants/ha.
- Moderate density: 250-500 plants/ha.
- High-density: 500-1250 plants/ha.
- Ultra-high-density: >1250 plants/ha.

  Meadow orcharding: In some European countries, 20000 to 70000 plants/ha are planted under super high-density or meadow orcharding, however, it has not been commercialized so far.
- Tree size control (key to successful HDP).

- ☆ Use size-controlling rootstocks (clonal rootstocks)/dwarfing rootstocks.
- ☆ Use spur-bearing scions.
- ☆ Training and pruning methods to induce dwarfing and spreading habits of trees.
- ☆ Mechanical devices like branch bending to control size.
- ☆ Use of chemicals/growth retardant.

Manures and Fertilizers

- ☆ Doses of manures and fertilizers depend upon soil fertility, soil type, kind and age of the tree, cultural practices, climate, crop load and rootstock. These should be applied on the basis of soil and leaf nutrient analysis.
- ☆ In an orchard of optimal fertility, N, P and K may be applied in the ratio of 70:35:70 g/year per plant, supplement this dose on a yearly basis and stabilize to 700:350:700 g NPK/tree after 10 years of planting. The application of FYM may also be supplemented with 10 kg per plant per year to the age of 10 years with a maximum of 100 Kg.
- ☆ Farmyard manure+P+K should be applied during December and January.
- ☆ Apply half N before bud break and the remaining half one month after flowering.
- ☆ A deficiency of calcium will lead to a physiological disorder in apple fruit called 'bitter pit' or 'cork spot'. Spray 0.5 per cent calcium chloride during the later stage of fruit development.
- ☆ Boron deficiency also causes internal browning in fruit and corking spots on the flesh (internal cork) and sometimes as cork tissue in the skin (external cork). Spray sodium borate @ 10 g/litre three times during flowering and fruit set will help to correct the disorder.

Irrigation

- ☆ Most of the apple orchards in India are situated in rainfed, sloppy areas.
- ☆ The most critical period for apple is from April to August. The peak water requirement is after the fruit set.
- ☆ Apple requires 114 cm of water per year, scheduled in 15 to 20 irrigations. During the summer and fruit development phase, irrigation is required at 7 to 10 days intervals and during the remaining period at 20 to 25 days intervals.
- ☆ Mulching and drip irrigation reduce the total water requirement of apples. Drip or trickle irrigation system has high water use efficiency in apples. It also suits the trees on hill slopes. Mulching can be done with straw, hay, sawdust or any other organic matter.

Training

- ☆ Trees are trained according to the growth habits and vigour of the plant. Training is done to give a desired shape as well as to get a canopy for intercepting maximum sunlight so that all the portions of the tree get sufficient sunlight and become productive.
- ☆ Standard trees are trained on a modified central leader system.
- ☆ Dwarf trees are trained on spindle bushes and slender spindles.
- ☆ The apple trees can be trained as a central leader, open centre, modified leader, cordon, dwarf pyramid, espalier and tatura trellis canopy.
- ☆ The apple bear fruit buds on previous season growth are called a spur.

Pruning

- ☆ Pruning in apples is essential to maintain a proper balance between vegetative growth and spur development during the dormant condition, *i.e.* January-February.
- ☆ Pruning is done according to the system of training followed.
- ☆ Apple bears fruit on spur as well as on tips of one-year old shoots.

Flowering

- ☆ The fruit buds are usually developed on spurs or on the tips of one-year old shoots.
- ☆ High relative humidity of more than 75 per cent in March delayed flowering and adversely affect the fruit set.
- ☆ Apple cultivars are diploid and triploid. The diploid cultivars are self-fruitful because they have plenty of good pollen. The triploid cultivars are self-unfruitful and have high female sterility. For these cultivars, pollinizers that produce plenty of viable and compatible pollen with commercial cultivars should be planted. The flowering period of the pollinizer should overlap with the commercial cultivars.
- ☆ For effective pollination, at least 6-8 bee hives/ha should be kept to increase fruit set in apple.

Fruit Thinning

- ☆ Heavy bearing in apples results in small-sized, poor-quality fruits and alternate-year bearing. Fruits can be thinned either by hand or chemically.
- ☆ Spray Ethephon 100-200 ppm at the petal fall stage results in optimal fruit thinning.

Fruit Drop

Most of the commercial cultivars experience three phases of fruit drop.

- **Early drop (natural):** Due to unpollinated and unfertilized flowers (Control: Add plant pollinizer cultivars).
- **June drop:** Due to fruit competition for nutrition and moisture stress. (Control: frequent irrigation and mulching).
- **Pre-harvest fruit drop:** It is severe in early ripening cultivars (40-60 %) and causes severe economic losses mainly due to reduced auxin levels or increased ethylene levels in fruit.

  Control: Application of 10 ppm NAA before expected fruit drop or 20-25 days before harvesting.

Harvesting

- Apple is a climacteric fruit, so harvest/pick the fruit at proper maturity.
- It has a non-uniform maturity of fruits, so more than one picking is required.
- Fruit stalk should remain intact with fruit at harvest. Handle the fruits carefully during harvesting for better shelf life.
- Precooling, grading and packing are required before marketing or storage.
- Apple has a long storage life, it varies from cultivar to cultivar. Most of the apple cultivars can be stored for 4 to 8 months at -1.1°C to 0°C with relative humidity of 85 to 90 per cent.
- Ambri has the longest storage life.
- Cider is a fermented wine prepared from apples.

Insects

- **Woolly apple aphid:** Woolly aphids are purple in colour and appear as cottony masses of their colonies. It weakened the tree by feeding on limbs and roots.

  Spray 0.2 per cent malathion or 0.2 per cent dimethoate. Use woolly aphid resistant rootstocks like Northern Spy and Robusta 5.
- **San Jose Scale:** It sucks the sap from leaves and branches, resulting in a decline of plants. In uncontrolled conditions, it may kill plants.

  Spray 0.2 per cent chlorpyriphos or dimethoate @ 0.2 per cent and 4 to 6 per cent diesel oil emulsion.
- **Blossom thrip:** It lays eggs by puncturing the developing fruit. The egg laying site develops into tiny rosette spots.

  Spary 0.05 per cent chlorpyriphos.
- **Apple root borer:** It is a serious pest of apple in the foothills of the Himalayas range. It feeds on living and dead roots.

  Drenching with 0.05 per cent chlorpyriphos.

- **Apple fruit moth:** Larvae insert into the fruit and feed its pulp. During severe attacks fruit drop off. Late varieties are more sensitive to the fruit moth.

  Bacillus thuringiensis is effective in controlling fruit moths.

Diseases

- **Apple scab:** It is a fungal disease that infects both leaves and fruit. Infected fruits are unfit for consumption and greenish to brown spots appear on the leaves. Follow clean cultivation, destroy all affected parts.

Table 16.1: Spray Schedule to Control Apple Scab

Spray No.	*Tree stage*	*Fungicide/100 litre of water*
1.	Silver tip to green tip	Mancozeb (400 g)/Captan 50 WP (400 g).
2.	Pink bud stage	Mancozeb (300 g).
3.	Fruit set (pea stage)	Captan 50 WP (300 g).
4.	Petal fall	Thiophanate methyl (50 g)/carbendazim (50 g).
5.	Fruit set	Mancozeb (300 g) and repeat at 14 days intervals.
6.	Pre-harvest spray (20-25 days before harvest).	Mancozeb (300 g).

- **Powdery mildew:** Powdery mildew infection starts from shoots, blossoms and fruit. It reduces the quality and yield of fruit.

  Spray wettable sulphur @ 0.02 per cent or carbendazim @ 0.2 %.
- **Root rot:** The cambium portion below the root bark turns brown in colour and may have a fermentation smell.

  Drenching with carbendazim @ 0.1 per cent.
- **Canker (pink disease, silver leaf canker, black rot, nail head canker):** A circular red spot appears on the stem bark and the size of these spots increases with severity.

  Drenching and spraying carbendazim @ 0.1 per cent.
- **Sclerotium blight in nursery:** It attacks near the base of the stem and its intensity increases with humidity.

  Drenching with 0. 3 per cent Thiram or 0.04 per cent aureofungin.
- **Crown gall:** It is caused by soil-borne bacteria. It enters the tree wound (training, pruning, cutting, damage, insect) and stimulates the hormone that causes the formation of a tumour or gall.

  Biocontrol with a related strain of bacteria (*Agrobacterium radiobacter* K-84). Prune the gall or remove a tree if infected heavily. Remove the soil surrounding the crown gall.

- ☆ **Collar rot:** Formation of dark brown, purple or grey canker spots on the bark of the lower portion of scion.

  Drenching with 0.3 per cent Mancozeb or 0.5 per cent copper oxychloride (COC).

Physiological Disorders

i) Scald: It is one of the storage disorders in apple. Light mottling on the greener surface of fruits are initial symptoms. Darkening becomes more severe with the elapsed time and ultimately extends to the red surface also. Scald usually affects the fruit skin, but in several cases, it may extend to fruit flesh. Immature fruits are more susceptible to scald, which is aggravated by warmer temperature during storage.

ii) Bitter pit: Characterized by small sunken spot on the fruit surface, which are more prevalent near the blossom end. Unlike the name, these corky tissues are never bitter in taste. Immature picked and large-sized fruits are most affected in the 'off' year. Golden Delicious, Yellow Newton and Gravenstein are the most susceptible cultivars.

Heavy doses of nitrogenous fertilizers, excessive shading and heavy pruning are reported to favour bitter pits. Storage at 32 to 34°F with 85-90 per cent relative humidity control pitting. Spray 0.5 per cent calcium chloride during the later stage of fruit development decreases the incidence of bitter pit.

iii) Internal browning: It is mainly associated with the cultivar Yellow Newton and characterized by brownish streaks radiating into flesh from the core. Controlled atmosphere (CA) storage with higher temperature can be helpful in controlling this disorder. The appearance of disorder is less at 1°C in cold storage.

Chapter 17

Pear (*Pyrus* spp.)

Family: Rosaceae; 2n = 34

Introduction

- *Pyrus communis*: European pear/soft pear.
- *Pyrus pyrifolia*: Japanese/Chinese/sand pear/hard pear.
- Among temperate fruits, pear is next to the apple in terms of area, production and varietal diversity. It is one of the important temperate fruit crops of India.
- India ranks 2nd in the world level in production of pear.
- It is primarily grown in hills at 1700-2400 m above sea level in Himachal Pradesh, Jammu & Kashmir, Uttarakhand and Uttar Pradesh.
- Low-chilling varieties of pear adapted well in the Subtropical Region of Haryana, Punjab, Assam and even in South India.
- Pear is a pome fruit and its edible portion is fleshy thalamus.

Origin

- *Pyrus* species are native to the Northern Hemisphere of the Old World.
- **Occidental group:** European and West Asian species (*Purus communis*) are native to Eastern Europe and South Western Asia.
- **Oriental group:** East and North Asian species (*Pyrus pyrifolia*) are native to Eastern Asia, including China, Japan.
- **Patharnakh pear (*Pyrus pyrifolia*):** Originated in China.

Soil and Climate

- This is one of the hardy temperate fruit crops grown in a wide range of soils from sandy to clay. However, medium textured, fertile, well-drained and deep soil without a hard pan up to 2 m are most suited. An ideal pH for this crop is ranged from 6.0 to 7.5. It can withstand more easily under wet soils than apples but is sensitive to drought conditions.

- Pear is a less chill unit requiring crop, and it can be grown on less elevation from sea level, *i.e.* about 1200-1800 m.
- Pear can be grown on a varied climatic condition, as it can tolerate as low as –26 °C during dormancy and as high as 45 °C during the growing period. However, temperature <-3.3 °C kills the opening blossoms.
- The soft pear (*P. communis*) cultivars require higher chilling hours (1000-1500 hrs) when grown on higher hills, whereas semisoft or hard pear (*P. pyrifolia*) require 200-300 chilling hours. Patharnakh pear is one of the low chilling hours requiring variety (150 chilling hours), which can survive easily in hot summer winds and high temperature.

Cultivars

On the basis of the origin of varieties, these are classified into three groups:

1. European types
2. Asian types
3. Hybrids of European and Asian types

However, on the basis of pulp texture (hardiness), the varieties are also classified into three groups:

- **Hard pear:** Patharnakh, Punjab Nakh.
- **Semisoft pears:** These are natural hybrids of *P. pyrifolia* and *P. communis* species *e.g.* Punjab Beauty, Le Conte, Baggugosha (*P. communis*), Punjab Gold, Keiffer, Punjab Nectar.
- **Soft pear:** Bertlett, Conference, Punjab Soft, Nijisseiki, Flemish Beauty.

Important cultivars

Patharnakh: It is a hard and heavy bearer pear. Keeping quality of fruit is good. Fruit skin is green in colour with prominent dots. The average fruit weight is 150 g.

Punjab Nakh: Fruit is hard and medium in size. The fruit's skin colour is greenish with prominent dots. Keeping quality is good.

Baggugosha: Tree is vigorous, upright and irregular bearer. Fruits are yellowish green in colour and tapering towards the stem end or bell shaped.

Punjab Beauty: Fruit is semi-soft and skin colour is yellowish with red blush. Flesh is white and juicy.

Punjab Soft: The fruit is soft and medium-sized. Skin colour is yellowish green. The flesh colour is white and suitable for processing purposes.

Bartlett: It is a prolific regular bearer. Fruits are medium sized, ovate shaped and skin turns yellow at ripening. Keeping quality of fruit is poor.

Rootstocks

- ***P. pyrifolia*:** Root suckers are used as rootstock.
- **Kainth (*P. pashia*):** Raised by seed after stratification.
- **Shiara (*P. serotina*):** Raised by seed after stratification.
- **Quince (*Cydonia oblonga*):** Clonal and dwarfing rootstock, but commercial cultivars of pear like Bartlett, Bosc, Eldorado need 'Old Home' as interstock as they are incompatibility with Quince. Quince B is semi vigorous and Quince C is dwarfing rootstock for pear.
- **Oregon 211:** Dwarfing, highly tolerant to wet soils, resistant to crown gall, collar rot, foot rot and root lesion nematodes.
- **Oregon 261 and 264:** Resistant to bacterial gummosis, canker crown gall and nematodes.
- **Oregon 249:** Dwarfing, resistant to crown gall, collar rot, powdery mildew and nematodes. Tolerant to high temperature, drought, high soil moisture content and low pH.

Propagation

- Pears are commercially propagated by shield or 'T' budding from June to August.
- In tongue and Cleft grafting, both stock and scion should be of the same size (January–February).
- Cutting: Used for cultivars as well as raising rootstocks (November – December).

Planting

- The appropriate time for planting is in the month of January during dormancy at 7 to 8 m apart.
- Fill the pit of 1 m^3 with the mixture of 50 cm top soil +50 kg FYM +SSP 500g.
- The hilly grown cultivars are partially self-fruitful, so they require pollinizer varieties. Pollinizers should be planted as 4th plant in every 4th row for better pollination and fruit setting.

Manures and Fertilizers

- Farmyard manure, SSP, MOP should be applied as a basal dose in the month of December.
- Apply ½ dose of urea before flowering in February and the remaining half in the month of April after the fruit set.

Table 7.1: Application of Manures and Fertilizers

Age of plant (year)	*Farmyard manure (kg)*	*Urea (g)*	*Single super phosphate (g)*	*Muriate of potash (g)*
1-3	10-20	100-300	200-600	150-450
4-6	30-40	400-600	800-1200	600-900
7-9	40-50	700-900	1400-1800	1050-1350
>10	50	1000	2000	1500

- Calcium deficiency in pear causes black ends and cork spots, so spray 0.5 per cent calcium chloride.
- Fruit cracking, corky tissue, calyx end rot and blossom blast in pear is due to boron deficiency, so apply boron @ 200 ppm.

Irrigation

- In hilly areas, it is grown as rainfed.
- Fruit development is a critical stage for the application of irrigation, *i.e.* during summer months. So, irrigate at 7-10 days intervals.
- Irrigation should not be applied from November to January.
- Shortage of water during the months of April, May and June restricts tree growth and leads to the formation of hard and small-sized fruits of poor quality.

Training and Pruning

- Trained with modified centre leader system.
- Heading back and thinning out should be followed in bearing trees.
- Pruning should be done keeping in mind that the flower occurs on two years old wood and spur for more than six years. So, no annual pruning is required. Only dead, dried, broken branches and limbs with a spur over six years are to be removed in the month of December.

Flowering

- Flowering starts in the last week of February and continues up to 3rd week of March.
- Flowers are borne on spurs and pollinated by insects (honey bees). However, honey bees prefer to pear flowers less due to the low amount of nectar.
- Fruit setting starts from the last week of March.
- Pear exhibits parthenocarpy and produces seedless fruits also.
- Fruit growth follows a sigmoid curve pattern.

Fruit Thinning

- Pear trees tend to overbear, which results in smaller sized fruits, so thinning should be done just after natural fruit drop during mid-April. One fruit per cluster may be retained.
- Foliar application of KNO_3 @ 1.5 per cent after full bloom improves fruit yield and quality of Patharnakh pear.

Harvesting

- It is a climacteric fruit.
- Fruit matures in 135-145 days after fruit set.
- Individual fruit may be harvested at maturity by giving a gentle twist rather than pulling to avoid spur damage.
- Fruits are pre-cooled at -0.6°C to -1.6°C to remove field heat and then ripened at 15.6 to 21°C with 80-85 per cent relative humidity.
- Pear can be stored for 4-8 months at -1.0°C and 85 per cent relative humidity.
- Core breakdown and scald disorder occur during storage.
- Perry is a wine prepared from pear.

Pear fruit can also be graded according to AGMARK grades as follows:

- Grading according to the diameter of the fruit:
- Extra large: 7.5 cm, Large: 5.6 cm, Medium: 5.5 cm, Small: 5.0 cm.

Insects

- **Hairy caterpillar:** It is a brownish coloured larva with a red head. It feeds on leaves and defoliates the plant.

  Collect and destroy the caterpillar.
- **Aphids:** Aphids suck the sap from leaves. Leaves become curly. It secretes honeydew like substances and falls on the fruits. The fruit becomes sticky, and black fungus may also develop on fruits.

  Spray 0.1 per cent dimethoate.
- **Bark-eating caterpillar:** It makes holes in the stem and raptures the tissue inside the stem, which blocks the supply of food material and water to the plant.

  Apply 0.1 per cent monocrotophos.
- **Fruit fly:** Fruit fly lay eggs in the young, tender and growing fruit. The larvae feed pulp and fruit drop prematurely.

  Spray 0.25 per cent fenvalerate.

Diseases

- **Pear scab:** Fruit becomes greenish to black at an early stage of development. In a later stage, a superficial spot appears on the fruit. It appears on leaves and twigs also. Spray 0.2 per cent Mancozeb and 0.2 per cent Bavistin.
- **Root rot:** Mycelium of the fungus grow under the bark of the root. It girdles at the crown and kills the plant.

  Spray 0.1 per cent Bavistin + 0.05 per cent Vitavax.
- **Shoot/fruit blight and bark canker:** Cankers develop sunken tissues. If the bark is cut above the sunken portion, reddish flecking can be seen on the wood. It also infects the shoot, blossom and rootstock.

  Spray Bordeaux mixture prepared in a ratio of 2:2:250 (2 kg copper sulphate, 2 kg unslaked/ quick lime in 250 litres of water).

Chapter 18

Peach [*Prunus persica* (L.) Batsch]

Family: Rosaceae; 2n = 16

Introduction

- Peach is the third most important temperate crop after apple and pear. It can also grown in the subtropical part of India. It became popular after the introduction of low-chilling varieties.
- Its production is 1.21 lakh MT in an area of 19000 ha.
- The peach become one of the popular temperate fruit crops due to its attractive colour, quality, taste and nutritive value.
- It is the lowest chilling unit requiring fruit among temperate fruits.
- It can be processed into several popular products such as preserve, jam, juice, frozen, dried and beverages. Nutritive food for babies can be prepared by mixing its pulp with milk.
- It is a rich source of vitamins, proteins, minerals and sugars. Its kernels are a good source of fat, protein, minerals and fibres.
- Peach seed contains 40 to 50 per cent oil.
- Important species of peach are *Prunus persica* var. *platycarpa, P. davidiana and P. mira.*
- The area under cultivation of this crop is in Uttarakhand (47.5 %), Punjab (29 %), Himachal Pradesh, Jammu & Kashmir, and Haryana.
- The main peach producing countries in the world are China, Italy, Greece, Spain, Turkey, USA, Iran, Chile, France and Argentina. China is the lead-

ing producer of peaches, accounting for about 62 per cent of the world's production.

- Peach fruit is botanically drupe. It is divided into two groups.
- **Cling stone peach:** In this type of peach, the pulp of the fruit remains clings to the stone.
- **Free stone peach:** In this type of fruit, the pulp detaches easily from the stone.

Nectarines

- It is *P. persica* var. *nucipersica,* which is a smooth-skinned peach fruit.
- The nectarine is a hairless or fuzzless peach and smooth-skinned. Fruits are smaller in size and somewhat like plum. Generally, the absence of fuzz on the nectarine is the only difference between nectarines and peach. Its pulp is like peach, but its outer portion appearance is like plum.

Origin

The peach is said to have originated from China and was found as early as 2000 B.C. This crop reached the Mediterranean region during very early from China. Thereafter, it spread throughout the world and reached India during the 19th century.

Soil and Climate

Deep, well-drained (free from water stagnation) loamy soils with rich organic matter are suitable for its cultivation. Heavy and fertile soils have more vegetative growth (chances of frost injury) and become susceptible to winter injury. Soils with pH between 5.8 and 6.8 are ideal. Soils with acidic and saline reactions are not suitable for peach cultivation.

The chilling requirement varies from cultivar to cultivar, *i.e.* 200-850 h. However, in most of the cultivars, the chilling unit requirement is 850 h. Low chilling peach requires less than 300 h of chilling requirement, and these can be grown in subtropical plains. The climate becomes warm to hot (35-42 °C) after fruit setting, *i.e.* from March to June. This climate is suitable for fruit development and maturity if sufficient irrigation/moisture is provided. It blooms 20 to 30 days earlier than apple and is susceptible to frost. To avoid the spring frost on blooms, apply 200 ppm GA_3 before the leaf falls. The orchard or area which is free from the early spring frost is suitable for peaches. The peach flowers in the early season and the bud swelling starts during early summer are injured by frost (-6.5 °C). The area of the deep valley is not suitable for peach production as the cold air settles in that area. The land having gentle slopes is ideal. It is prone to the water-logged and sun-scald.

Varieties

Most of the peach cultivars are self-fruitful except a male sterile variety J.H Hale.

- **Red Haven (Hale aven x Kalhaven):** It is an early maturing stone free cultivar. Fruits are medium-sized. Skin colour is yellowish to red and attractive.
- **Alexander:** It is early season maturing variety. Fruits are medium sized, round and free from stone, but keeping quality is low. Fruit's skin colour is beet root purple.
- **July Elberta:** It has medium to large-sized fruits and grown commercially. Fruit skin colour is pale yellow to red and is good for canning purposes.
- **Shan-i-Punjab:** It is a low chilling variety suitable for subtropical regions. Fruit maturity time is the first week of July.
- **Prabhat:** It is low chilling and medium to heavy bearer variety. The flesh is yellow to red firm and slightly acidic.
- Other important varieties are Elberta, J.H. Hale, Stark Red Gold.

The peach varieties are classified on the basis of use: i) Table cultivars, ii) Canning cultivars.

Table Cultivars

The cultivars are free from stone with yellow-coloured flesh and regular bearer *e.g.* Alexander, J.H. Hale, Elberta, Cardinal *etc.*

Canning Cultivars

- Yellow fleshed.
- Cling stone with non-splitting pit.
- Uniformity in size and maturity of fruits *e.g.* Certex, Crawfords Early, Golden Bush, Halford, Fortuna *etc.*

Nectarine Cultivars

- The peach which is free from hairs/fuzz.
- Smooth-skinned nectarine cultivars are used for table purposes *e.g.* Nectarine, Sunlite, Sun grand, Sun Red, Sun Ripe and Sun Rise.

Male Serile Cultivars

J.H. Hale, June Elbarta and Halberta.

Low Chilling Cultivars

Partap, Florda Prince, Florida Sun, Shan-e- Punjab, Sun Red, Safeda, Matchless, Sharbati, Parbhat, Khurmani, Early Grande.

Propagation

Most commonly, it is propagated by grafting and budding.

Cutting

Peach can be planted during November-December by cutting. For better rooting, cutting should be treated with 1000 ppm IBA, first to dissolved in 50 per cent alcohol. Plant these cuttings on a raised bed. The rooting success rate of IBA treated cuttings is about 50-80 per cent, whereas the success rate of untreated cutting is 30 per cent.

Grafting

Tongue grafting (rootstock and scion of the same size) and wedge grafting (rootstock is thicker) during January–February.

Budding

T- budding in the months of June-July.

Rootstock

- ✰ The peach species grown wild are vigorous and strong, so they are used as rootstocks.
- ✰ Peach seedlings as well as cultivars like Sharbati, Khurmani and Flordaguard, are generally used as rootstocks.
- ✰ Plum, apricot, almond seedlings and sometimes behmi (*P. mira*) can also be used as rootstocks.
- ✰ Pazza (*Prunus padus*): Commercially used rootstock in Himachal Pradesh.
- ✰ Nemaguard, GF 557, Yunnan, Rubira, Nemared, Flordaguard and Shalil are nematode-resistant rootstocks of peach.
- ✰ Montelar: Resistant to 'Fe' and 'Mg' deficiency.
- ✰ GF 677: Vigorous rootstock, tolerant to dry, wet and high pH soils.
- ✰ Siberian Sea: It is cold hardy rootstock.
- ✰ Seed stratification at 4-10°C for 10-12 weeks in moist sand.
- ✰ Pre-sowing treatment with thiourea @ 0.5 per cent, GA_3 @ 200 ppm or BA @ 100 ppm reduces the stratification period and improves seed germination and seedling growth.
- ✰ Myran: Suitable for poor soils. Tolerant to drought, root-knot nematode and *Verticillium* wilt.

Planting

Planting of peach is done during the winter season. The planting in the hilly areas is done by contour or terrace method of planting, however, in plain areas square method of planting is followed. Pits of 1 m x 1 m x 1 m size at a spacing of 6 m x 6 m in hills are dug during September-October. Planting is done in the month of December-January (Bare rooted), however, with earthball in the month of July-August. Preference for planting should be given during the winter season. In high-density planting, spacing can be reduced to 3 m x 3 m. In tatura trellis system of high-density planting 6 m x 1 m spacing is practised.

Manures and Fertilizers

Manures and fertilizers should be applied by broadcasting in the tree basin and mix them thoroughly up to 15 cm depth under the canopy of the plant.

Table 18.1: Application of Manures and Fertilizers

Age of plant (year)	*Farmyard manure (kg)*	*Urea (g)*	*Single super phosphate (g)*	*Muriate of potash (g)*
1	10	225	190	150
2	15	450	380	300
3	20	675	570	450
4	25	900	760	600

The basal dose of farmyard manure, along with single super phosphate and muriate of potash should be applied in the months of December- January during the dormant period. Nitrogen should be applied in two split doses, 1st in the spring season (February- March) or before flowering and 2nd dose, *i.e.* the remaining half after the fruit set. This crop is sensitive to a deficiency of iron. It can be supplied by foliar application of $FeSO_4$ @ 0.5 per cent.

Irrigation

Sufficient moisture is required in the root zone of the plant before initiation of flowering, during the fruit development stage or in hot summer. Moisture deficiency causes fruit drop, shrivelling of fruit and decreases the quality of fruits. At least two irrigations must be given during fruit development. Irrigation should not be given during dormancy, *i.e.* October-January, to allow the plant to harden during cold weather or just before harvesting of fruit.

Training and Pruning

Peach is a temperate fruit, so light is one of the limiting factors. To provide sufficient light to the canopy, a modified or open centre system of training is to be followed. The flower bears on the one-year old shoot once in a lifetime, so heavy regular pruning of 1/3rd of all shoots is required.

The important component of pruning is heading back and thinning out. Pruning is required in such a way that it produces about 25-50 cm annual growth to maintain optimum productivity.

Flowering and Fruiting

Flowering starts in the first week of February and continues till the end of February. Sites which are free from spring frost during bloom are most suitable for early season peach. To avoid frost injury during flowering, apply GA_3 @ 200 ppm or ethephon before dormancy or before shedding of leaf to delay flowering.

Peach flowers are pollinated by insects. Most of the peach varieties are self-fruitful except J.H. Hale, which requires a pollinizer variety for pollination. Fruit setting starts during the March. Patterns of increase in length, diameter and fruit weight follow a double sigmoid curve. Fruits are borne on one-year old growth.

Inter-culture

Inter-culture operation should be done regularly, but ploughing should not be done more than 10 cm deep. Most commonly, it is required during the winter season after harvesting of crops.

Intercropping

The short duration leguminous or non-exhaust crops can be grown as intercrop up to 3-4 years of planting or before the start of fruiting *e.g.* Soybean, cowpea, pineapple and turmeric. Exhaust crops like onion and okra should be avoided.

Crop Regulation

Excessive flowering and fruiting reduce the quality of fruits, size of fruits and reduce the flowering and fruiting in the ensuing season. For the production of good quality fruit, crop regulation is to be followed. The thinning can be decided on the basis of the following criteria:

- Fruits to leaf ratio (25 to 30 leaves per fruit is appropriate).
- Regulate the flowering time.
- Space between fruits (10-15 cm).
- Number of fruits per tree (on the basis of the vigour of tree).
- Expected fruit size.
- The number of fruits to be retained depends upon.

  Fertility level of the orchard.

  Tree vigour (capacity of the plant).

  Cultivar (varies from cultivar to cultivar).

Harvesting

- The plant starts bearing after 2- years of planting. It is a climacteric fruit, so harvest the fruit at the appropriate stage, *i.e.* during June- July, and it gives an economic yield up to 20 years. The average yield from fully grown trees varies from 50 to 125 kg in hilly areas.
- Harvest fruit in 3 to 4 pickings at 4 days intervals. For distant marketing, harvest the hard fruit that has attained good colour. However, for the local market, harvest ripe fruit by twisting with hand.
- Harvest the ripe fruits with good flavour and aroma at a temperature of more than 15°C. It remains undesirable at 10°C, however, at 4.4°C, fruits break down internally instead of ripening.

- ☆ Peach can be stored for 2-3 weeks in cold storage at 0 to 3.3°C and relative humidity 85-90 per cent.

The maturity of fruits is decided on the basis of following maturity indices:

Computation Method

- ☆ Days taken to flowering or flowering to fruiting.
- ☆ Days to blooms are as under:
  - 76 days in Partap.
  - 83 days in Shan-e-Punjab.
  - 86 days in Alexander.
  - 127 days in Elberta.

Physical Method

- ☆ Fruit size.
- ☆ Firmness of fruit.
- ☆ Colour development.
- ☆ Pitch discolouration.

Chemical Analysis

- ☆ TSS (total soluble solids).
- ☆ Acidity of fruit (%).
- ☆ Sugars in fruit (%).
- ☆ Sugars: acid ratio.
- ☆ Starch content.

Grading and Packaging

- ☆ Before packaging, sort the injured, damaged and undersized fruits from the lot and grade them according to size and stage of maturity.
- ☆ Special grade: ≥ 55 mm, Grade I: 46-55 mm, Grade II: <46 mm. Generally, 2 and 4 kg CFB are used for packaging of fruits.

Insects

Peach leaf curl aphid: Nymph and adult suck the cell sap, leaf becomes curled and drops down in severe cases. It also affects flowering and fruiting.

Introduction of *Aphidius colemani* as a parasite of this aphid. Important predators include lady beetles, soldier beetles, green lacewings, brown lacewings and syrphid flies.

Peach fruit fly: Female lays eggs on mature fruit, larvae/ maggot feeds pulp and fruit becomes unfit for consumption.

Grow early maturing cultivars like Shan-e-Punjab, Florida Sun, Partap, Parbhat. Deep ploughing in May-July and spray dimethoate @ 0.1 per cent.

Diseases

Shot hole: It occurs on twigs and buds. Small purple-black lesions appear on the twigs and dark brown scattered lesions on leaves. These spots enlarge with light colour in the centre and a purple brown colour in the margins.

Apply Captan or Ziram or Thiram @ 0.2 per cent.

Brown rot: It infects shoots, blossoms and fruit. The infected blossom turns brown and drops down. Canker formation occurs on shoots and these may dry.

Spray Captan @ 0.2 per cent.

Physiological Disorder

i) **Sun scald:** Constant exposure to the bright sun causes sun scald on the trunk, twigs and branches. Paint the exposed area with lime or wrap straw or hay around the stem.

ii) **Splitting:** It occurs at the pit hardening stage and gum exudates from the fruit make it unfit for consumption. The gumming and splitting occur due to heavy rains after a long drought spell. However, the exact reason is still not known.

Production Technologies for Cultivation of Minor Fruits

Chapter 19

Pineapple (*Ananas comosus* (L.) Merr.)

Family: Bromeliaceae, 2n = 50

Introduction

- It is a Crassulacean acid Metabolism (CAM), monocotyledon, monocarpic and herbaceous fruit plant.
- This fruit was reached in India by the Portuguese in 1548.
- The wild Brazilian pineapple (*A. microstachys*) is considered to be the ancestor.
- It is an important commercial tropical fruit of the world. This is also known as "Golden Queen".
- It can be grown as a mixed crop, as a hedge plant, or as a sole/pure crop.
- It is tolerant to drought and can be grown on marginal lands.
- It is an excellent source of vitamin 'C' and also a good source of vitamins 'A' and 'B'.
- Its fruit contains 'Bromelin', a protein that acts as a digestive enzyme.
- The edible part of fruits is called coalesced berries.
- India ranks 2nd in area and 6th in production of pineapple in the world.
- Pineapple is suitable for high rainfall, humid coastal region and peninsular India. The major pineapple growing states are West Bengal (20.2 %), Assam (17.4 %), Karnataka, Meghalaya, Manipur, Nagaland, Tripura, Bihar, Arunachal Pradesh, Mizoram, Kerala, Goa *etc.*

- It is grown in countries such as Tropical America, Malaysia, Mexico, Taiwan, Kenya, Sri Lanka, Australia, Singapore and India.

Origin

- The origin of this fruit is believed to be Brazil.

Soil and Climate

- Plants grow well in all types of soils; however, heavy clay soils are not desirable.
- It performs well in sandy loam soil. Soil should be 50-60 cm deep without a hard pan or stones. Avoid low laying areas with high water tables.
- It prefers acidic (pH 5.0-6.0) soils. Alkaline soils are not considered suitable for pineapple cultivation.
- High Mg and Mn content is detrimental to the crop.
- It is a crop of tropical regions, so grown in the coastal regions. It can also be grown in subtropical areas where the temperature remains optimum from 22 °C to 32 °C.
- The proper growth of leaves takes place at 32 °C, and roots grow well at 29 °C. The plant growth ceases at temperature below 20 °C and above 36 °C. It is sensitive to frost.
- High night temperature is harmful, but the desirable temperature difference between day and night is 4°C. A mildly cool winter improves fruit quality.

Cultivars

- **Kew/ Giant Kew:** It is a commercially grown late-maturing variety suitable for canning purposes. Fully ripe fruits are yellowish in colour and fruit weight is 2 to 3 kg.
- **Queen:** It is cultivated for table and canning purposes. Mature fruits are golden yellow in colour. Fruit weight is 1 to 1.5 kg.
- **Mauritius:** Fruits are medium-sized, oblong-shaped and have yellowish to deep red skin.
- **Jaldhup:** It belongs to the Queen group. Fruits are suitable for table and breeding purposes. It has an alcoholic flavour.
- **Lakhat:** This cultivar belongs to the Queen group. Fruits are suitable for table and breeding purposes.

Other important varieties are Charlotte, Rothschilds, Cayenne and Red Spanish.

Propagation

- It is propagated vegetatively by suckers, slips and crowns. Suckers and slips are usually preferred for planting as they flower comparatively earlier than the crowns. Using uniformly sized planting material of monotypes is required for uniform crops.

- **Suckers:** Suckers arise from the axils of the leaves on the main stem. Suckers are of two types, sprout from the stem just below the ground level (ground suckers), and aerial sucker arises from the main stem above the ground level.
- Suckers takes about 15 to 18 months to come into flowering. Suckers of about 450 g weight are the best planting material.
- **Slips:** It sprouts on the stalk just below the fruit, it takes 20-22 months to flower after planting. A slip weight of 350 g is best for planting.
- **Crown:** About 40-45 cm crowns are selected for planting. It takes 24 months to flower.
- **Leaf cuttings:** During a shortage of planting material, the leaf cuttings from the crowns can be used for multiplication.
- **Micro-propagation:** It is the fastest method for multiplication of plants.

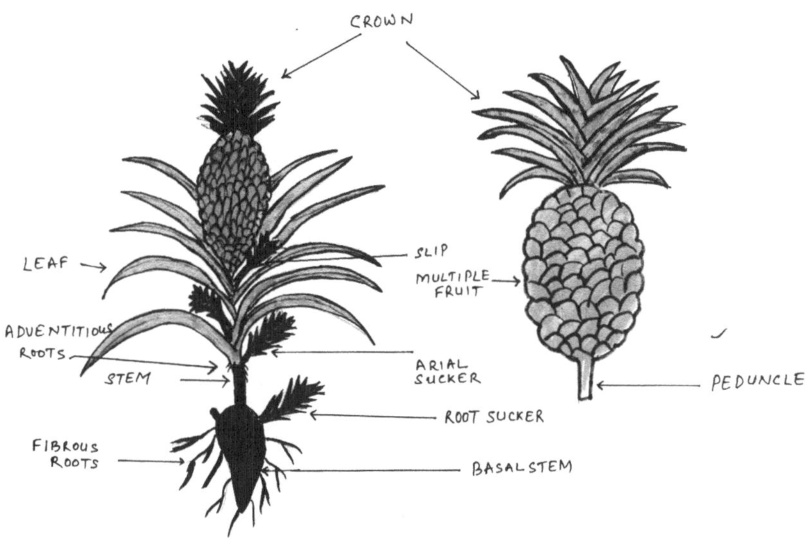

Fig. 19.1: Sketch of Pineapple Plant

Planting

- Planting should be done 12-15 months before the natural flowering season and it varies with the area of cultivation. Planting is generally done during *Kharif* season (July-August). Planting time for different areas is as under:
    - Assam: August-October planting.
    - Kerala/Karnataka: April-June planting.
    - West Bengal: October-November and June-July planting.
- Pineapple planting can be done in single or double-row systems.

- There are four methods of planting *viz.,* flatbed, furrow, contour and trench.
- Planting intensity varies with the area of cultivation.

  In a subtropical area, 63400 plants/ha at a spacing of 22.5 cm x 60 cm x 75 cm.

  In hot and humid areas 53300, plants/ha at a spacing of 25 cm x 60 cm x 90 cm.

  In rainfed, highly fertile, hilly areas, 43500 plants/ha at spacing 30 cm x 60 cm x 90 cm.
- High-density planting is profitable with other advantages like fewer weeds, protection from sunburn (leaves grow upright, resulting in the natural covering of fruits), increased production of suckers and slips per unit area and avoiding lodging of plants.

Manures and Fertilizers

- Pineapple is a shallow feeder or high nutrient requiring (low depth) crop.
- Pineapple is a heavy feeder crop. The nutrient requirement of this crop varies with the density of planting per unit area.

  Apply @ FYM 30 t/ha/year.

  NPK @ 500:140:560 kg/hectare for 42,000.

  NPK @ 600:200:600 kg/ha for 64000 plants/ha.
- Apply fertilizers 3 months after planting in 4 split doses. Apply nitrogen as foliar application @ 2 to 4 per cent.

Micronutrients Deficiency

- Micronutrient deficiencies such as Fe, Zn and Cu are common in pineapple. Their deficiency can be fulfilled by the application of following chemicals:

  $FeSO_4$ @ 0.5 to 1.0 per cent as foliar application.

  $ZnSO_4$ @ 0.5 to 1.0 per cent as foliar application.

  $CuSO_4$ @ 1.0 per cent as foliar application.

Cultural Operations

- **Irrigation:** Pineapple is mostly grown under rainfed conditions, supplementary irrigation will help in the production of good sized fruits. In hot months, 4 to 6 irrigations during the dry season at 20 to 25 days intervals to ensure good crop. Mulching is helpful in moisture conservation.
- **Earthing up:** It is a most important cultural operation in pineapple, it provides good anchorage to plants and prevents them from lodging as its roots are shallow.

Remove Suckers, Slips and Crown

- **Suckers:** Suckers start growing with the emergence of inflorescence. Fruit weight increased with increase in number of suckers per plant. Hence, desuckering should be delayed as much as possible.
- **Slips:** Slips grow with the developing fruits. Increased number of slips delayed fruit maturity. So, slips should be removed as early as they attain plantable size.
- **Crown:** The crown has no effect on the fruits. So, crown removal is not required as it makes the better appeal of fruit, but handling of fruits becomes difficult.

Use of Growth Regulators

- Spray NAA (planofix) @ 10-20 ppm to induces flowering.
- Spray ethrel @100 ppm or ethrel @ 25 ppm + urea @ 2 per cent + $NaCO_3$ or $CaCO_3$ @ 0.04 per cent to induce flowering.
- 35-40 functional leaves are required for flowering and to produce normal sized fruit.
- Application of NAA @ 200-300 ppm at 2-3 months after fruit set increased fruit size by 15-20 per cent.
- Application of ethrel @ 500 ppm before harvesting helps in uniform ripening and development of colour, but fruit becomes acidic and results in a lack of flavour.

Ratoon Crop

De-suckers the plants immediately after harvesting of fruits leaving one sucker on the mother plant and all slips should be removed. Never allow more than two suckers. Add fertilizer and earthing up for better anchorage of the ratoon crop.

Weed Control

Pre-emergent application of Bromacil and Diuron @ 2 kg a.i/ha is effective in controlling weed.

Flowering and Fruit Thinning

- Pineapple plants generally start flowering 11 to 12 months after planting.
- Irregular flowering in pineapple is a major drawback. Flowering in pineapple may be induced by keeping stems in a horizontal position due to geotropism.
- Pineapple exhibits parthenocarpy and self-incompatibility. Self-incompatibility is a major breeding hindrance in pineapple.
- Spray NAA (planofix) @ 10 to 20 ppm to induce flowering.

- Application of ethrel @ 25 ppm along with 2 per cent urea and 0.04 per cent sodium carbonate induces flowering.
- Fruit takes 4½ to 5½ months from fruit set to harvest.
- The fruit growth of pineapple follows a sigmoid curve.

Harvesting

- Flowering takes place in 10 to 12 months after planting, generally in the month of February to April and ripening in the months of July to September (15 to 18 months after planting).
- Fruits that mature in winter are more acidic and not fit for table purposes.
- It is a non-climacteric fruit, so it should be harvested along with a 5-7 cm stalk after attaining maturity.
- Fruits harvest at slight colour change at the base of the fruit for canning purposes.
- For table purposes, harvest fruits at the golden yellow colour stage.
- Fruit can be stored for 10 to 15 days after harvest in ordinary conditions and 5-6 weeks at temperature 10 to 13°C with 85 to 90 per cent relative humidity.

Insects

- **Mealybug:** It is a polyphagous pest that damages more than 100 plant genera. It feeds on leaves (green area) and fruit. It weakens the plant and decreases the quality of fruits.

  Use healthy suckers. Prune and dispose of the infested shoots and apply neem oil. Do not apply water and fertilizer in excess amounts.
- **Thrips:** Thrips feed on crown leaves and develop centric ring patterns on fruit, resulting in the formation of dead eye spots.

  Spray malathion @ 0.1 per cent.

Diseases

- **Soft rot, storage rot and fruit rot**: Formation of water spots on the stalk end of the fruit. Later on, these fruits produce a foul smell.

  Apply 10 per cent benzoic acid in alcohol at the cut end of the fruit.

  Avoid to collect suckers from infected plants or areas. Destroy the diseased plants.
- **Heart rot and stem rot (*Phytophthora parasitica*)**: Water soaked spots appear on the basal part of leaves in the whorl. A central whorl of the leaves is affected. It spreads on the chlorotic leaf area and converts the leaf from olive green to brown.

  Apply Captan @ 0.1 per cent.

- **Leaf spot:** Grey or yellow coloured spots appear on the leaves. In the later stage, leaves have a dried spot.

  Use healthy planting material and apply Trichoderma as a biocontrol agent.

  Sclerotium wilt: The infected leaves turn yellow-red to bright red from the leaf tip and it spreads downward with time.

  Mealybug acts as a vector, so control mealybug.

Physiological Disorders

i) Sun scald: Caused due to direct fall of sun rays on the exposed area of fruit and results in damage of fruit cells under the sun-exposed skin. Its surface gets damaged and causes the rotting of fruits. Covered the fruits with dry straw or banana leaves or its own leaves during lodging and leaning in April - May.

ii) Multiple crowns: Normally, pineapple bears one crown, but sometimes it can bear up to 25 crowns. So, the fruit tip becomes flat and broad with an insipid taste. It may be due to the application of excessive nitrogen. Sometimes due to environmental and soil factors. This disorder is mainly found in the Cayenne group (Kew).

iii) Black heart: The base of the fruitlets shows brownish spots and later on black discolouration of the centre core. This disorder is induced due to low temperature or excessive use of gibberellic acid (GA). Exposing the cold stored fruits to high temperature (40°C) for 24 hours reduces black heat to some extent.

Chapter 20

Pomegranate (*Punica granatum* L.)

Family: Lythraceae (earlier it was Punicaceae); 2n = 16,18

Introduction

- ☆ National Research Centre for Pomegranate is at Kegaon, Solapur, Maharashtra.
- ☆ It is a nutritive fruit with high medicinal value. Hardy nature, low maintenance cost, high yields, better-keeping quality, less water requirement and good market value make cultivation of this crop profitable.
- ☆ India ranks first in the world with respect to area and production of pomegranate. India contributes 70 per cent of the world's pomegranate market.
- ☆ Depending on growing conditions, pomegranates have both deciduous and evergreen natures.
- ☆ Under temperate conditions, it is deciduous, while under tropical and subtropical conditions, it is evergreen.
- ☆ Pomegranate is a Latin name of fruit that means 'grainy apple'.
- ☆ It is a hardy and bushy plant with a tendency to develop multistems.
- ☆ In India, it is cultivated in 2.7 lakh ha with a production of 30.86 lakh MT.
- ☆ It is grown commercially in Maharashtra (63 %), followed by Gujarat (16.2 %), Karnataka, Andhra Pradesh, Madhya Pradesh, Telangana, Tamil Nadu, Rajasthan, Uttar Pradesh *etc.*
- ☆ It is mainly grown in Mediterranean regions such as Iran, Morocco, Spain and Egypt *etc.*

- About 40 per cent of fruits are exported from India to UAE, Holland, UK, Saudi Arabia *etc.*

Origin

- It is native to Iran (Persia).

Soil and Climate

- It can be grown on diverse types of soils such as limy, alkaline and saline and also thrives well in shallow rocky gravel soils having pH 5.5 to 7.5. It is rated as a salt hardy plant and can tolerate salinity up to 9 dSm^{-1}.
- It performs better in well-drained sandy loam to deep loam soils.
- Colour development and quality of fruit are lower in heavy textured soils as compared to light soils.
- It can be grown in tropical to warm climates, however, it is a subtropical crop.
- It grows best in semi-arid climates with cool winters and hot dry summers with low rainfall.
- The plant performs better under dry conditions also, but for good yield and quality, irrigated conditions are the best.
- The hot and dry conditions are required during fruit development and ripening.
- It is a winter hardy tree but injured when the temperature goes below -11°C.

Cultivars

- **Ganesh (GBG1):** It is a revolutionary cultivar for pomegranate. Selection from Alandi cultivar and developed by Dr GS Cheema at Pune in 1936, and it was renamed Ganesh in 1970.
- **Alandi:** It is commonly grown in Maharastra and Gujarat.
- **Dholka:** Large-sized fruits with a greenish white rind.
- **Kandhari:** Big-sized with deep red seeds.
- **Jalore Seedless:** Soft-seeded variety developed from CAZRI, Jodhpur, for arid regions.
- **Mridula:** Seedling selection from F_2 population of a cross between Ganesh x Gul-e-Shah Red.
- **PAU Selection:** Chance seedling of Ganesh. The arils are white transparent with a little pink tinge, but the seeds are little hard.
- **Basin Seedless:** Red-coloured arils, soft seeded and grown in Karnataka.
- **Bhagawa (G-137):** It is the ruling variety and clonal selection from Ganesh. Fruits are medium to large and red in colour.
- **Kabul:** Big-sized fruits with a yellow coloured rind.

- **Muscat Red:** Medium-thick skin.
- **Paper Shell:** It is grown in South India.
- **Jyoti:** It is open-pollinated seedling of Bassein Seedless and Dholka developed by UAS, Bangalore.
- **Jodhpur Red:** Grown in Rajasthan.
- **Ruby (Ganesh x Kabul x Yercard):** Developed by IIHR, Bangalore. The fruits are similar to Ganesh.
- **CAZRI Vishal (Ganesh x Khog):** Early maturing, soft-seeded cultivar with a yellowish-red skin colour.
- **Phule Anardana:** Arils are bold, acidic and have a blood red colour.
- **Amlidana:** Ganesh x Nanha.

Propagation Methods

- **Stem cutting:** Hardwood stem cutting can be rooted satisfactorily from December to January when the plant shed its leaves, or during September-October. For early rooting, cutting should be soaked in 100 ppm IBA for 24 hours.
- **Air-layering:** Propagated by air-layering only in the high humid areas during the rainy season. Treat the uppercut portion with IBA @ 10000 ppm.
- **Tissue culture:** It is also standardized for commercial propagation.

Planting

- Planting at 5 m x 5 m spacing (400 plants per hectare). Planting time is January-February or July to August.
- Dig a pit of 60 cm x 60 cm x 60 cm; fill it with 25 kg FYM and 1 kg SSP along with the top 50 per cent soil. However, for high-density planting, 5 m x 2 m spacing (1000 plants per hectare) gave 2-3 times higher yield and profit without affecting quality as compared to normal planting.
- It was also recommended that for an initial 4 to 5 years, to get a higher yield, a distance of 5 m x 2 m may be adopted, thereafter, alternate plants may be removed to maintain a planting distance of 5 m x 4 m.
- Ganesh can be planted at 3 m x 3 m and Kandhari at 4 m x 4 m apart.

Manures and Fertilizers

It is a hardy plant that grows well in low fertile soils. Productivity can be increased by fertilizer application.

Apply 25 kg farmyard manure, 1.0 kg urea, 1.5 kg single super phosphate and 600 g MOP. Farmyard manure should be applied in the month of December. Apply ½ Urea + SSP + MOP in February and remaining ½ urea in April.

Irrigation

- Irrigate the orchard regularly to avoid cracking and better development of fruits. Irrigation after fertilizer application at 7 to 10 days intervals during summer and 15 to 20 days intervals during winter is required.
- Drip irrigation is best to reduce fruit cracking and fruit drop in addition to harvest higher yield of good quality fruits. It saves 43 per cent of water and increases yield by 30-35 per cent.

Training

- The plants can be trained as single or multi-stemmed. A multi-stemmed training system is followed in India.

Pruning

- Planting time is from December to January.
- No regular pruning is required. Removal of suckers, dead or diseased branches and crisscross shoots for better development of the framework.
- High-growing shoots are headed back to produce laterals for better fruiting.
- Remove the main stem by heading back after 10 years of planting to allow side shoots for flowering and fruiting.

Flowering and Fruit Setting

The rest period is required for the plant to enhance yield during a particular period. It is provided by withholding irrigation for about 60 days in advance of the normal flowering, and roots are exposed in bahar treatment. Pomegranate has two types of flowers: bell-shaped (male flower) and vase-shaped (hermaphrodite).

There are 3 main seasons (bahar) of flowering in western India:

1. **Ambe Bahar:** Flowering in February –March. This is followed in an area where a sufficient amount of water for irrigation is available because the fruit development stage coincides with the dry period/summer season. Fruits are available from July to August and shade leaves from November-December.
2. **Mrig Bahar:** Flowering starts during June-July. This crop is taken in the area where a sufficient amount of irrigation water is not available during hot weather. Fruit ripens in October and continues up to December. Flowers are forced to meet water requirements during the rainy season. Withhold irrigation from December to April, which suppresses its growth and causes shedding of leaves in March-April. Apply fertilizer and light irrigation, followed by two heavy irrigations at 7 days intervals before rain set. There will be profuse growth within 15 days, along with flowering and fruiting.

3. **Hasta Bahar:** Generally, it is not taken as it is risky and less economical.

Harvesting

- The tree starts bearing from the 3rd or 4th year onward, but the full economic yield is achieved after the age of 6th to 7th years.
- Pomegranate fruits become ready for harvesting in 120 to 130 days after fruit set.
- It is a non-climacteric fruit harvested at a fully ripe stage.
- At maturity, the fruit gave a metallic sound on tapping and crunching sound when pressed.
- Seeds dried with pulp are called anardana.
- About 16-20 kg fruits, or 80-120 fruits per plant can be harvested from full-grown trees.

Grading

The rates in the market are on the basis of the grading of the fruits. The grading is done on the basis of rind colour, size and weight.

Super sized: >750 g, King sized: 500-750 g

Queen sized: 400-500 g, Prince sized: 300-400 g

Fruits can be stored for longer periods (5-6 months) at 4.5°C and relative humidity 80-85 per cent.

Insects

- **Anar fruit fly/fruit borer (*Deudorix isocrates*):** It is a serious pest that makes holes in the fruit and excreta of insects emerge from the hole. Affected fruits will rot and fall down.

  Cover the fruits with paper or cloth or, plastic bags. Destroy affected fruits.
- **Bark-eating caterpillar (*Indarbela tetraonis*):** Same as in Aonla.
- **Mealybug:** It attacks whole plant parts and is found more on leaves, buds and fruits. Insects look like cottony growth in the infested areas. Cover the single stem plant with an alkathene sheet.

  Spray 1.5 litre quinalphos (Ekalux) in 500 litre water.
- **Stem borer (*Coelosterna spinator*):** Grub enter into the stem by making a hole and feeding inside the hole, which blocks the supply of nutrients and water. Control is the same as in bark eating caterpillars.

Diseases

- **Bacterial leaf spot:** Irregular water-soaked spots on leaves, flowers and fruits, which resulted in premature defoliation. It also feeds on stems and branches.

Streptocycline 200 ppm +0.1 per cent copper oxychloride.

- **Cercospora leaf spot:** Affected plant shows brown coloured circular spots on leaves and fruits later on, these become irregular in shape.

  Spray 0.2 per cent copper oxychloride.
- **Alternaria leaf spot:** Appears small sized round spots of brown colour on leaves and fruits. In the advanced stage, patches become larger and the fruit starts rotting.

  Spray Mancozeb (Indofil M-45) @ 0.2 per cent.
- **Fruit rot (*Phomopsis* sp.):** In fruit rot, fungus grows inside the fruit without external symptoms. Arils get infected and fruits are unfit for consumption. In the advanced stage, brown spots on fruits appear.

  Spray 0.2 per cent copper oxychloride.

Physiological Disorder

Fruit Cracking or Splitting

The fruit cracking may be due to the following reasons.

- Moisture stress due to irregular irrigation or drought conditions followed by heavy rain or heavy irrigation.
- High temperature, hot and dry winds.
- Deficiency of Boron, Potash and Calcium.

Fruit cracking is more common in Ambe Bahar (spring crop) under North Indian conditions.

The outer skin or peel of the fruit becomes hard due to long dry spells and rainfall or irrigation after the dry spell, the pulp grows but the peel becomes plastic or doesn't allow the pulp to grow or take more size.

Control Measures

- Planting of tolerant cultivars like Bedana, Jalore Seedless and Khog.
- Apply regular irrigation and mulching to maintain moisture.
- Early harvesting of fruits immediately after maturity.
- Spray Borax @ 0.2 per cent to meet the boron deficiency.
- Plant windbreak around the pomegranate plantation.

Chapter 21

Jackfruit (*Artocarpus heterophyllus* Lam.)

Family: Moraceae; 2n = 56

Introduction

- It is a tropical fruit grown in high-rainfall, highly humid and coastal areas of the world. It is a favourite fruit of many, owing to its sweetness.
- The fleshy carpels, which are botanically the perianth are edible portions. The flavour is ethyl butyrate. Most commonly, it is used as table purpose fruit, apart from this, it is also used in processing for preparation of pickles, jack leather and *papad*.
- Seeds of this fruit are a rich source of starch and protein, and are delicious.
- Jackfruit is also called Monkey Jack.
- Wood has timber value and is used for the preparation of different furniture.
- The dried leaves are used for the preparation of disposable plates.
- Its fruit is the source of complete nutrition for the consumers.
- Its fruits are as healthy as avocado and olive in terms of nutrients for human beings. It has almost all the nutrients equivalent to a mother's milk.
- It is rich in vitamins B and C; calcium, potassium, iron, proteins and carbohydrates. It is easily available as well as affordable.

- The fruit has "Jacalin" content, which is helpful to prevent cancer and AIDS.
- In our country, the trees are distributed in southern states like Kerala, Tamil Nadu, Karnataka, Goa, coastal Maharashtra and other states like Assam, Bihar, Tripura, Uttar Pradesh and the foothills of the Himalayas.
- The jackfruit tree is widely cultivated in tropical regions of India, Bangladesh, Nepal, Sri Lanka, Vietnam, Thailand, Malaysia, Indonesia and Philippines.

Table 21.1: Nutritive Value of Jackfruit (100 g)

Constituent	*Average value*
Moisture (%)	76.20
Energy (cal)	88.00
Protein (g)	1.90
Fat (g)	0.10
Fibre (g)	1.10
Carbohydrates (g)	19.80
Potassium (mg)	107.00
Calcium (mg)	20.00
Phosphorous (mg)	41.00
Iron (mg)	0.56
β Carotene (mg)	175.00
Thiamine (mg)	0.03
Riboflavin (mg)	0.13
Niacin (mg)	0.40
Vitamin C (mg)	7.00

Health Benefits of Jackfruit

- **Strengthens immune system:** It is a good source of Vitamin C, so it protects from viral and bacterial infections.
- **Protects against cancer:** In addition to vitamin C, jackfruit is also rich in phyto-nutrients such as lignans, isoflavones and saponin, which have anti-cancer and anti-ageing properties.
- **Aids in healthy digestion:** Jackfruit is also known to contain *anti-ulcer* properties which help to cure ulcers and digestive disorders. In addition to the presence of high fibre, it prevents constipation and helps in smooth bowel movements.
- **Maintains healthy eyes and skin:** Jackfruit contains vitamin A, a powerful nutrient known to maintain healthy eyes and skin.

- ☆ **Boosts energy:** It is a good source of simple sugars like sucrose and glucose, which provides energy or booster energy.
- ☆ **Lowers high blood pressure:** It is a good source of potassium, which is helpful in lowering blood pressure and reduces the chances of heart attack and strokes.
- ☆ **Control asthma:** The extract of boiled roots of jackfruit helps to protect from asthma.
- ☆ **Strengthens the bone:** It is a good source of Mg, which is helpful in the absorption of calcium and the strengthening of bones. It also protects from bone related disorders.
- ☆ **Prevents anaemia:** Being a good source of iron, it helps in proper blood circulation in the body and prevents anaemia.
- ☆ **Maintains a healthy thyroid:** It is a good source of copper, which is helpful in thyroid metabolism, and production of hormones and its absorption.

Origin

India is considered a native of jackfruit.

Soil and Climate

i. It gives good yield in warm humid areas of hill slopes and hot humid areas of plains.

ii. It grows well from sea level to 1200 m above sea level.

iii. The optimum temperature requirement of the crop is 22 to 35 °C.

iv. High humidity areas of the west coast plane is found highly suitable. However, performance under low humidity is medium in terms of yield and quality.

v. It can be grown on a wide range of soils but open textured loam or deep rich alluvial or red laterite soils having slightly acidic reactions (pH 6.0 – 6.5) are ideal for its cultivation.

Varieties

Based on the firmness of the bulb cultivated or naturally found jack trees are broadly classified into two types *viz.*, firm flesh (kappa) and soft flesh (rassal). A brief description of some important varieties is given below.

- ☆ **Konkan Prolific:** The tree is monoecious with small male flowers held by a pedicel. Female flowers are larger than the male and the pedicel is thick. Fruits are multiple type, medium sized immature (8 to 9 kg) fruit's skin colour is green and ripe fruit is greenish yellow to brownish yellow.

- **Singapore (or) Ceylon Jack:** It was introduced into Tamil Nadu from Sri Lanka. Fruits are medium in size (7-10 kg). It is a precocious bearer and seedling progenies will start bearing from 3 years after planting. Fruits are available from March to June and again from September to December.
- **Hybrid Jack:** It is a cross between Singapore Jack x Velippala developed at Fruit Research Station, Kallar, Tamil Nadu. Trees are precocious in bearing and carpels are bigger in size and sweeter than the parents.
- **Burliar-1 (T Nagar Selection):** It was developed at Fruit Research Station, Burliar, Tamil Nadu. The trees are medium in height and prolific bearer.
- **PLR-1 (Palur 1):** It is a high-yielding variety developed at Vegetable Research Station, Palur (Tamil Nadu Agricultural University). The fully ripe fruits have flat stigmatic surfaces instead of spiny surfaces. The special feature of this variety is that the trees bear fruits twice a year *viz.*, March to June and October to December. The average fruit weight is 12 kg and it contains 115-120 bulbs.
- **PPI-1 (Pechiparai 1):** It was developed at the Horticultural Research Station of Tamil Nadu Agricultural University at Pechiparai by clonal selection from Mulagumoodu local. Tree bears in two seasons, *viz.*, April – June and November – December.
- **Muttam Varikka:** It bears medium-sized fruits with crispy sweet flakes.
- **Gumless Jack:** It bears medium sized round fruits with crisp flakes and the fruit is low in latex.
- **Other cultivars:** Sindoor, Siddu, Shankara

Propagation

- **Seed**: Seeds are sown just after extraction from fruit and they lose their viability during storage. The seed treated with NAA @ 25 ppm for 24 hours improves germination and growth of seedlings.
- **Air layering:** Vegetative propagation can be done by air layering the cut portion should be treated with IBA @ 500 ppm for more success.
- **Epicotyl grafting/Softwood grafting:** Sowing of jackfruit seeds during the 1st week of July in polythene bags of 10 cm x 20 cm size. When the seedlings attained 12-14 cm height these were used as rootstock. Select scions of the previous season growth about 4 cm from the terminal portion. Insert the scion in rootstock and tie it with polythene strips.

Planting

- For planting, 1 m x 1 m x 1 m pits are dug. The upper half of the soil should be mixed with 50 kg farmyard manure, 500 g single super phosphate, 50 ml chlorpyriphos and filled in the pits.

- ☆ The grafts are planted in the centre of the pits during June-September. Proper staking is required to avoid lodging and subsequent breakage at the graft joint. Thatching or providing shade with coconut leaves and thorough mulching of basins after the cessation of monsoon season is essential to avoid the mortality of grafts.
- ☆ Seedlings are planted from June to August in a square system of planting.

Manures and Fertilizers

A fully mature bearing tree requires about 20-30 kg of farmyard manure, 200 g nitrogen, 320 g P_2O_5, 960 g K_2O and 5 kg ash. The plant should be irrigated just after application of manures and fertilizers. These should be applied under the canopy of the plant about 50-60 cm away from the main trunk.

The manure and fertilizers should be applied during monsoon season, and if irrigation facilities are available, these can be split into two doses, *i.e.* during June –July and September–October.

Irrigation

This crop can be grown under rainfed conditions but performs better under irrigated conditions. The requirement for irrigation depends upon soil type, growth stage of the plant and season. The crops should not be under water/moisture stress conditions, especially during flowering and fruit sets. However, accessive irrigation leads to an insipid taste. The young growth stage of the plant is more sensitive to drought. Proper seedling growth cannot be achieved under stress conditions.

Training

Jackfruit should be trained to a single stem at least up to 1.5 to 2.0 m and then allow side shoots. The flowers in jackfruit appear on the main stem, so vegetative growth should not be allowed on the main stem.

The bearing of flowers from the main stem and scaffold branches is called 'cauliflorous'. The flower bud on the main stem should be protected from the stray animals.

Inter-cultivation

To utilize the interspace between the jackfruit plantation, short-duration crops (legumes and vegetables) can be grown during the initial phase or up to the gestation period of the crop. It gave extra income to the farmer.

Harvesting

Seedling plants of jackfruits start yielding from 7 to 8 years of planting, however, vegetatively propagated or grafted plants start yielding from 3rd year of planting. In Singapore, the seedling jack plants start yielding from the third year of planting.

In plain areas, the fruits are available from March to June, however, in hilly areas the availability of fruits extends up to September. Some genotypes bear fruits during the offseason and their fruits are available during October to December. The average fruit weight is 10 to 30 kg and the number of fruits per plant varies from 20 to 100.

Insects

- **Shoot and trunk borer:** The caterpillar makes a hole into the shoot bud and damages the fruit. The infected part should be removed.
- **Brown weevil (*Ochyromera artocarpi*):** It makes holes into the tender buds, shoots and fruits.

  Collect and destroy the fallen fruits and buds. Spray monocrotophos @ 0.035 per cent.

Diseases

- **Fruit rot (*Rhizopus artocarpi*):** It causes premature fall of young fruits due to rotting and may result in heavy loss of yield under highly humid conditions.

  Spraying Dithane M-45 (0.2 %) or Bavistin (0.05 %) or Fytolan (0.2 %) at 15 days intervals during fruit growth.

Chapter 22

Strawberry (*Fragaria x ananassa* Duch.)

Family: Rosaceae; 2n = 56

Introduction

- Strawberry is grown in different parts of the world, and this is a heart-shaped fruit of love, which was mentioned by the Roman Poets Virgil and Ovid in the first century. In England, gardeners had cultivated strawberries since the sixteenth century.
- Over the last few years, strawberries have experienced one of the highest rates of consumption growth. It is the most consumed fresh fruit, rich in proteins, minerals Ca, P and K, and a fair source of vitamins A, B_1, B_2, niacin and vitamin C. It is regarded as one of the best natural sources of antioxidants.
- Besides, strawberries can be processed for making wine, jam, jelly, ice cream, soft drinks *etc.*
- The word strawberry comes from the farmer's practice of covering the berry plant using straw.
- The United States is the world's largest producer of strawberry and most of the strawberry industries are located in California.
- The United States produces almost 28 per cent of the world's supply. Spain accounted for 8 per cent, followed by Russia, Korea, Japan and Poland accounting for 5 to 6 per cent each.
- The strawberry plant is a herbaceous perennial rosette that has a compressed stem called the crown.

- ☆ The stem or crowns have a central pitch core and can survive for several years because of secondary cambium activities.
- ☆ The leaves are compound, pinnate and trifoliate with leaflets that vary in shape from wedge to ovate.
- ☆ The axillary meristem may develop into stolon's or branch crowns, with a regular positioning along the axis and in strict relation to the crown growth rate.
- ☆ In India, major strawberry producing states are Haryana (58 %), Mizoram (20 %), Meghalaya, Kerala and hilly tracts of Himachal Pradesh, Uttaranchal, parts of Uttar Pradesh and Kashmir valley. However, its cultivation has spread to tropical and subtropical zones, where other states like Meghalaya, Sikkim and Mizoram have taken up the cultivation of this high-value crop.

Table 22.1: Nutritive Value of Strawberry Fruits (per 100 g of edible portion)

Components	*Values*
Energy (Calories)	44
Water (%)	87-90
Fat (%)	0.2-0.5
Protein (%)	0.23-0.70
Total sugars (%)	3.3-9.1
Fibre (%)	1.1
Vitamin A (IU)	60
Vitamin C (mg)	26-120
Calcium (mg)	21
Phosphorus (mg)	21
Potassium (mg)	164

Origin

Strawberry originated in Europe during the 18th Century.

Soil and Climate

- ☆ The strawberry plant grows the best in sandy loam soil; on very light soil, frequent irrigation is needed for the establishment of runners.
- ☆ The heavy soils with sufficient soil moisture are better than light soils.
- ☆ Strawberries prefer soils reasonably rich in humus because 70-90 per cent of their roots are found in the top 15 cm of soil. Avoid growing continuously on the same piece of land and on the land previously devoted to potato, tomato, eggplant, pepper and raspberry because these crops are hosts for *Verticillium* fungus.

- The plant can be grown in soil having a pH of 5.0 to 7.5. However, it performs better in soil pH of 5.5 to 6.5.
- Strawberry can be grown up to an elevation of 3000 meters above mean sea level. In humid or dry regions, it is widely grown under protected cultivation and open conditions in temperate and subtropical countries with a maximum temperature of 22 to 25°C during the day and 7 to 13°C during the night.
- Among all the different types of berries, strawberry gives the quickest return in the shortest possible period.
- During the winter season, frost and winter injuries are harmful to plant growth.
- Frost may damage the centre of the open flowering, causing the characteristic black eye.
- The frost injury may be reduced using mulch, covering the row with plastic and creating good air drainage.
- During vegetative and flowering phases, changes in photoperiod, vary the plant morphology and yield of strawberries.
- Stolon formation, petiole length, leaf area and yield increase with the increase in photoperiod.

Cultivars

Table 22.2: Important Cultivars of Strawberry

Sr. No.	*Cultivars*	*Fruit weight (g)*	*Yield per plant (g)*	*Total soluble solids (°B)*	*Ascorbic acid (mg/ 100 g pulp)*
1	Ofra	12.0	228	14.0	107.0
2	Camarosa	17.0	248	10.2	70.7
3	Sweet Charlie	14.2	165	9.6	68.0
4	Chandler	12.5	187	11.2	87.2
5	Elista	7.9	175	9.4	79.3

Other cultivars

- **Jutogh Special:** Producing more runners.
- **Shimla Delicious:** It is late season cultivar and fruits are delicious.

Soil

The most suitable medium is a 1:1:2 ratio of soil: sand: farmyard manure.

Site Selection

Prime importance factors for selection of site for commercial strawberry plantations are:

- ✰ Appropriate climate.
- ✰ Supply of good quality water.
- ✰ Proper slope and better soil type with good drainage facility.
- ✰ Accessibility to market.
- ✰ Transportation facility.
- ✰ Adequate labour availability.

Propagation

- ✰ Strawberries can be propagated through seed, vegetative means, as well as through tissue culture.
- ✰ Propagation through seed is not advisable because they do not produce true-to-type plants. A single berry produces thousands of seeds. Seeds require stratification to break the dormancy. The effective chilling temperature for breaking dormancy ranges from -2 to 6.5°C for 2 to 4 weeks.
- ✰ Strawberry is propagated vegetatively by runners. The stolen are arises from the leaf axil as creeping stalk during summer months.
- ✰ Initially, the runners have small fibrous roots, these are detached after attaining sufficient growth for planting as a new plant.

Runner Production

- ✰ Although runner produces true-to-type plants, but viral diseases are quite often transmitted through runners only. So, a separate bed should be used for runner production.
- ✰ During the selection of a site for runner production, it should be kept in mind that strawberry had not been grown on the same piece for the last 3 to 4 years. The planting should be done at 1.2 m x 1.2 m or 1.8 m x 1.8 m distance. All the cultural practices should be strictly followed.
- ✰ Production of 30 to 40 runners per plant is considered a satisfactory runner production rate.
- ✰ The production of runners can be enhanced with the application of GA_3 @ 50-100 ppm.
- ✰ For good growth and more survival, the separated runner should be transplanted in a polybag having a 1:1:2 ratio of sand: soil and FYM for one month.

Cold Storage of Runners

- ☆ Runners can be kept in cold storage for planting during spring.
- ☆ Runners are lifted during late November-December and stored until June-July.
- ☆ Runners should be packed in bundles of 200-250 in polythene bags, not more than 0.07 mm thickness.
- ☆ Runners can be kept in good condition for 3 months at 0 or –1 °C.
- ☆ In North Indian conditions, these runners are planted in September-October.
- ☆ Runners should be removed from cold storage during April- May and planted as soon as possible to avoid injury from heat.

Land Preparation and Planting

- ☆ The site selected for strawberry cultivation should be prepared thoroughly and kept free from weeds.
- ☆ Mix farmyard manure on the topsoil and avoid mixing it in deep soil because top soil application is more effective.
- ☆ Fumigate the soil with chloropicrin (33 %) before planting to check the population of nematode, verticillium wilt and weeds. Follow crop rotation to avoid soil-borne diseases.
- ☆ Healthy runners with well-developed root systems should be planted in the field.
- ☆ The success rate or survival rate of runners depends upon the depth of planting and pack of soil around the roots. A crown may die due to more depth or too shallow planting.

Time of Planting

During summer in hilly areas and during winter (*Rabi*) in plain areas (North India).

Growing in Containers

- ☆ Container cropping is very attractive for strawberries because of better phytosanitary conditions, fruit picking facilities and different techniques of production, however, in open conditions, cultivation is possible only in a normal season.
- ☆ Strawberries can be planted in containers and can be shifted according to climatic requirements, *e.g.* in shade/sun or protected structure.
- ☆ For container gardening, a layer of drainage materials (hardcore, broken bricks *etc.*) are filled and then cover the drainage layer with potting compost to the level of the lowest ring of holes.

- For pot cultures of strawberries, soilless composts or those with low loam content porous media are recommended.

Cultivation under Plastic Low Tunnels

High temperature, long photoperiod and rainfall are unfavourable for strawberry cultivation. However, the off-season demand for high quality fresh strawberry remains high between March and August because there is a lean period for other fruits in the market.

In fact, growing strawberries out of their normal season has led to the emergence of the idea of protected cultivation of strawberries.

Protected cultivation basically involves the construction of either plastic greenhouses or polythene tunnels or row covers of plastic film over the wire hoops for forcing and prolong harvesting season by protecting plants from the vagaries of adverse weather. Dwarf varieties that grow better under protected conditions are Elista, Fair Fox, Gorella, Douglas, Fern, Selva and Chandler.

Advantages of Protected Cultivation

- Better growth and development of plants.
- Early crop maturity, early flowering by 10-15 days and fruits ready for harvest a week earlier.
- During a prolonged harvest period, normally fruit size starts decreasing after the second picking, but under protected conditions, growers can get a normal size up to 4th picking.
- Off-season availability of fruit.
- The chances of frost injuries are eliminated due to the rise in temperature inside.
- Low incidence of soil-borne diseases.
- Increase yield and improve the quality of fruits.
- Easy runner production in high rainfall areas.

Tips for Cultivation in Low Tunnels

- Soil fumigation.
- Use substrate media rich in humus.
- Use pre-emergence weedicide to check the weeds.
- Use drip irrigation along with fertigation.
- Frequent foliar sprays rather than top dressing.
- Use pest and disease-resistant dwarf varieties.
- Provision of honeybees for pollination.
- Provide vents/remove plastic from time to time for ventilation.

- Runners should be lifted and transplanted in poly tubes one month before transplanting in tunnels.
- Harvesting at alternate days because the rise in temperature accelerates ripening faster.
- Don't allow runners on the main bed.

Irrigation

- Irrigation is essential just after transplanting for rooting of runners.
- It is useful for the utilization of applied nutrients in soil, which prevents frost damage during the blossom and picking period.
- Frequent irrigation rather than a few irrigations.
- Heavy irrigation causes the incidence of Botrytis rot.
- To avoid excessive moisture/water, irrigation should be applied through drip irrigation.

Following points should be kept in mind while irrigating strawberries.

- Strawberry needs about one inch of water per week during the growing season.
- Apply water before the appearance of water deficiency symptoms.
- It is a shallow-rooted crop, so apply frequent and light irrigation.
- Sufficient irrigation/water is required during flowering and fruit development up to the fruit colour change stage.
- Irrigation just before harvesting reduces the TSS and shelf life of the fruits.

Weed Management

Keep the field free from weeds, it becomes expensive under wet weather conditions. Apply weedicides pendimethalin @330 g/l to keep the weeds under control.

Mulching

The objectives of mulching are as under:

- Maintain soil temperature.
- Reduce weed intensity.
- Maintain moisture level.
- Reduce soil erosion.
- The most commonly used mulching materials are straw, marsh hay, pea vine and sawdust.
- The time of mulching is also important, very early or late mulching is deleterious to the plant.

- Mulching with a black polythene sheet increases yield, advancement of crop and check weed germination.
- Plastic mulching sheets should be removed after harvesting the crop and disposed of properly.

Special Cultural Practices

- **Bud and shoot thinning:** Removal of 1-2 buds per plant improves fruit yield and quality; however, excessive bud removal reduces the number of fruits and drastic yield reduction.
- **Deblossoming:** Deblossoming is the removal of flowers to prevent fruiting, especially in runner production beds. Removal of blossoms from mother plants increases the production of early saleable runners.
- **Remove runners:** The runners should be allowed to root along the rows until sufficient crowns are formed, but the excess runners should be removed from the rows.
- **Intercropping:** Strawberry is sometimes planted as intercrops in young orchards. The strawberry grows very quickly and starts fruiting within 120 days after planting and continues up to three years if managed properly. It provides additional income till the establishment of the orchard.

Fig. 22.1: Strawberry Cultivation

Harvesting/ Picking

- Strawberries are harvested at different stages of ripening, depending on the time and distance to market.

- ☆ For local market, fruit should be picked at the pink colour stage or 3/4th colour development stage and collect the picked fruits into the crates for marketing.
- ☆ Fruits are harvested in shallow trays, which are highly perishable and get damaged easily if bulky containers are used.
- ☆ Berries are picked along with the caps or calyx.
- ☆ Special stem-grade fruit are picked along with the stalk, one to two inches from the calyx.
- ☆ Berries showing rot, sunburn, insect injury, severe misshape, soft and over-ripe should be discarded.
- ☆ Crates should not be filled full, berries may get crushed while stacking.

Post-harvest Handling

- ☆ Being a highly perishable fruit, the storage of fresh strawberries, even for short periods is not recommended.
- ☆ However, the berries must be transported to market and thus may require protection for at least 1 to 6 days from harvest.
- ☆ During transportation, it ripens very fast and get spoiled very easily.
- ☆ To improve the post-harvest life and marketability, careful handling, precooling and packing are to be followed carefully.
- ☆ Remove all fruits showing signs of decay from the plant during each picking.
- ☆ Do not put decayed fruit into the crate.
- ☆ Apply fungicides in the field to reduce decay in the field and storage.
- ☆ Precooling the fruit as soon as possible after harvest by keeping these in a cold place. Lower the temperature, slower the advancement of decaying organisms. For best control, keep the fruits as close to 0°C as possible.

Packaging

- ☆ It is a very perishable fruit, so it is not advised to store it even for a short period.
- ☆ It is a climacteric fruit, so ripened fruits get spoiled at a very fast rate.
- ☆ The strawberries are packed in plastic punnets and place in corrugated fibre trays or ventilated cardboard boxes.
- ☆ Store these corrugated fibre boxes and punnets-filled trays in the shade to avoid transpiration losses.
- ☆ Wrap strawberries with semi-permeable film immediately after harvest, it minimize the weight loss of the fruits and results in the formation of a carbon dioxide enriched atmosphere.

- Refrigerate the fruits at 5°C or below; if the fruit is supposed to keep for more than one day, reduce the temperature to 0°C.

Insects

- **White grubs and cut-worms:** It is 'C' shaped, white coloured insects with a brown head.

  It cut the underground portion of the plant. Deep ploughing and soil treatment with insecticidal dust and drenching the soil with chlorpyriphos 0.1 per cent during March is the remedial measure.
- **Hairy caterpillar:** These insects defoliate the plant.

  Spray malathion 0.05 per cent. Harvest berries at least 7-10 days after spray.
- **Red spider mite:** It reaps the surface of the leaf and feeds chlorophyll. During the later stage, it causes the mottling of leaves.

  Paired application of Abamectin at 7 to 10 days intervals. Neem oil is also effective to control it. Other insecticides like chlorpyriphos and carbamates may also be used.
- **Root weevil:** Larvae feed on the roots results in poor yield and stunted growth of the plant.

  Apply parathion (0.017 %) around the plants to reduce pest population and damage.
- **Thrips:** Larvae and adults feed on the fruit surface, resulting in the formation of netting. It decreases the appearance and shelf life of fruit.

  Deltamethrin @ 2.25 g a.i./ha reduces thrips incidence and the number of distorted fruits.
- **Leaf roller:** Larvae feeds on the leaves and roll them. In severe cases, it reduces the fruit size.

  Remove dead vegetation and spray *Bacillus thuringiensis* or entrust formulation of spinosad.
- **Tarnished plant bug:** Nymphs feed on seeds of fruit and adults damage the fruits.

  Dimethoate (0.5 %) is recommended when bugs are active at the first bloom stage and repeat after ten days.
- **Slugs:** They feed on the fruits and make rough holes in ripe fruits. It also feed on leaves. Control weeds, removal of excessive mulch, and planting at lower densities may reduce slug infestations. Use the stale beer slug trap.

  Prevent through cultural management, including the incorporation of trickle irrigation.

Diseases

- **Verticillium wilt:** The leaf margin turns reddish yellow to dark brown, which results in droop or wilt of older and outer leaves.

  Follow crop rotation, fumigate the soil with Formalin (5000 l/ha) or apply chloropicrin (210 l/ha).

- **Powdery mildew:** It infects leaves, flowers and fruit. In heavy infection, there is no fruit formation, and if fruit formation is there, then these are deformed.

  A spray of Bayleton or Karathane (0.05 %) at 15 day intervals between May to September is quite effective to control the disease.

- **Red stele:** It occurs in low-lying areas or poorly drained soil. Plants become stunted and may die in severe conditions.

  Soil drenching with disinfectant (Ridomil) and treat the planting materials before planting with Bordeaux mixture (1.0 %) or copper oxychloride (0.1 %) for 15 to 20 minutes to reduce the occurrence of the red stele in the field.

- **Leaf spot:** Superficial black spots appear on the leaf margin and the centre of the leaf remains green during moist weather.

  Two to three sprays of Bordeaux mixture (copper sulphate 800 g+ lime 800 g in 100 l water) at 10 to 15 days intervals along with one spray of carbendazim (l00 g/200 litre water) or Bayleton (2 g/litre) is effective.

- **Leaf scorch:** Occurrence of many small-sized purple spots on the outward leaf surface.

  Control the leaf scorch as per leaf spot.

- **Leaf blight:** It damages the plant through destruction of older foliage. In severe cases, the plant becomes defoliated and results in death of the plant.

  Control measures for leaf spots and leaf scorch will also control leaf blight.

- **Grey mould:** Fungus excrete the substance that kills the plant.

  Mulching with black polythene/paddy straw before flowering and spray with carbendazim/thiophanate methyl (100 g) or Captan 50 WP (600 g) just after blossoms open and repeat after 15 days.

- **Viral diseases:** It cause the yellowing edge of young leaves, which become crinkled.

  The control measure includes the use of a virus-free runner, maintaining isolation distance from infected plants, control virus spreading vectors by application of systemic insecticides and grow resistant or tolerant varieties.

Production Technologies for Cultivation of Nut and Plantation Crops

Chapter 23

Coconut (*Cocos nucifera* L.)

Family: Arecaceae/Palmae, 2n = 32

Introduction

- *Coconut*: Central Plantation Crop Research Institute, Kasargod, Kerala.
- Genus *Cocos* is monotypic, containing only one species, *i.e. Cocos nucifera*. The generic name '*Cocus*' is derived from the Spanish word Coco, which means monkey face. The coconut palm, with its tall cylinder and uniformly thick stem, massive crown, with large number of leaves, bearing bunches of nuts in their axils, is one of the most beautiful and useful trees in the world.
- Coconut is one of the important palms as it supplies food, drink (coconut water), shelter as well as raw materials for several industries.
- Almost all the parts of this tree are used in one way or another for a mankind, so it is regarded as the tree of heaven (Kalpa Vriksha). The monolinuron content of coconut oil has anti-HIV properties. An endocarp is used for making toys and buttons.
- India, Indonesia and the Philippines are leading countries in coconut production. The annual production in India is 13.92 m MT from an area of 2.15 m ha. The productivity of coconut is 7608 nuts per hectare in India, which is the best in the world. In India, the maximum productivity is in Maharashtra (20621 nuts/ha).
- Under normal conditions, every leaf axil of bearing palm can produce 12-15 spadix or inflorescence. The inflorescence generally develops within a strong top, pointed double sheath called a spathe. The coconut palm is monoecious.

- In Indian conditions, female flower production during March-May is higher but lower during September-January. Generally, the number of female flower production per inflorescence varies from 10-50 and these remain receptive for 19 to 20 days after the opening of the spathe.
- Generally, the Dwarf palms are considered autogamous (*i.e.* direct self-pollinated), while tall palms are allogamous (cross-pollinated). Both winds and insects are considered to be the main pollinating agents. After fertilization, it takes 11 to 12 months for the development of a mature nut.
- Most of the female flowers/buttons are unable to develop into fruit, only 2 to 40 per cent of flowers develop into fruits or reach maturity. Less flower to fruit ratio is due to the following reasons:

  Lack of pollination or fertilization.

  Defect in female flowers.

  Due to some physiological disorders.

  Due to insect pests and diseases.

  Unfavourable environmental conditions.
- Four month old nuts have the maximum amount of good quality refreshing coconut water. The meat formation begins at the age of 5 months in the form of a thin jelly like layer inside the shell or endocarp. The nut attains full size, after attaining the age of seven months, the water inside the shell starts converting into meat and the endocarp also becomes hard.
- When the meat is 11 to 12 months old, the shell, becomes fully hardened and the fresh weight of the meat is maximum at this time. Its fruit is a single-seeded drupe, where endosperm forms the edible portion.
- Major coconut growing states in India are Kerala (33.5 %), Tamil Nadu (26.2 %), Karnataka (23.7 %), Andhra Pradesh, Orissa and West Bengal *etc.*
- India ranks 3rd in the world in area as well as production of coconut.

Origin

On the basis of evidences available on the growing area of the coconut, it is found somewhere in South East Asia in the countries Malaysia or Indonesia. However, the exact origin is still a mystery.

Soil and Climate

It can grow well under a wide range of soil conditions. It does well in light and easily permeable soil. It can, however, be grown on heavier soils with good drainage. Coconut roots are very thick and unable to penetrate compact soil and the fluctuating water table also restricts root development. High rainfall areas, deep loamy and rich in plant food (nutrients) are ideal soil for its cultivation. This crop requires deep, well-drained, loamy soils for better production, however, it can be grown in all types of soils ranging from light sandy soils to heavy soils with a pH range from 5.2 to 8.0. Water deficiency during summer hinders the nut production to a greater extent.

About 90 per cent of the area under coconut production lies in tropical regions in the latitude between 20°N and 20°S. The optimum temperature for growth is 27°C. Palm requires plenty of sunlight and does not grow well under shade. It is extensively cultivated up to an altitude of 600 m above mean sea level. It is a crop of humid tropical areas. The well-distributed annual rainfall between 1300-2300 mm is optimum for its growth. The growth of palms in terms of palm girth, fruit setting and fruit/nut development depends on rainfall to a greater extent. However, heavy rainfall over a shorter period prevents pollination.

On the other hand, a severe drought at the flowering stage may kill the growing point and result in no fruit set.

Propagation

Palm is a perennial crop that exhibits considerable genetic variation and is mainly propagated by seeds. Horizontal placement of the nut is better than vertical placement for germination. In the propagation of palms, the only method of propagation is through nuts and the selection of nuts plays an important role. The nut should be selected from a garden with a record of consistently high yield.

Criteria for Selection of Nut

1. For the production of a nuts, the shape of the crown should be spherical or drooping, and an erect crown should be avoided.
2. The length of bunch stock and petiole should be short and stout.
3. The size of the nut should be medium and spherical to round in shape.
4. The barren nut-producing tree should be avoided.
5. The age of palm plants should be in the range of 25-60 years.

Collection of Seed Nuts

- The time of harvest varies from region to region.
- The age of nuts should be about 11 to 12 months for seed nuts.
- On the West Coast, the nuts for seed should be harvested during February-May.

Planting of Nuts

One to two months of storage of seed nuts is beneficial before planting. The seed nuts could be stored in polythene bags up to 8 months and during storage, 63-73 per cent of seed nuts start sprouting. These seed nuts could be planted horizontally or vertically with stock end up. However, horizontal planting has been found to be the best.

Cultivars

There are two types of cultivars, *i.e.* Tall and Dwarf. Natural crossing is responsible for a wide variability among different cultivars. The dwarf cultivars are Lakshadweep, Maldive Dwarf, Andaman Dwarf, Malayan Green Dwarf, Chowghat Orange Dwarf and Ganga Bondam. Whereas tall cultivars are Fiji, S.S. Green, San Roman, Philippines Ordinary, East Coast Tall, West Coast Tall.

- ☆ **West Coast Tall:** It is the most popular cultivar of India, mainly grown on the west coast. It starts yielding in 6-8 years, and on average, it produces 60-80 nuts/palm/year. The nut contains an average copra weight of 156 g with an oil content of 72 per cent by weight.
- ☆ **Lakshadweep:** Native cultivars of Lakshadweep Island, all the characters are similar to West Coast but yield heavily up to 120 nuts/palm/annum.
- ☆ **Andaman Ordinary**: This is a tall and massive cultivar largely cultivated in Andaman Island. The nuts are fairly large with an average copra weight of 173 g and low oil content yield (66 per cent by weight basis) but are resistant to some of the commonly occurring diseases.
- ☆ **Kappadam**: The nuts are large with a mean copra weight 285 g/nut, producing 60 nut/palm.
- ☆ **Leguna**: High-yielding type, widely grown in the Philippines and mature nut weight is 1.25 kg.
- ☆ **San Roman**: It is a cultivar of Philippines and the nuts are larger than Leguna and are heavy yielders.
- ☆ **Ganga Bondam**: Semi-tall type, grown in certain tracts of Andhra Pradesh, the nuts are medium-sized with high content of quality copra and the copra has up to 72 per cent oil and produces 65 nuts/palm/year.

Hybrids

- ☆ **Chandra Sankara:** Chawghat Orange Dwarf x West Coast Tall.
- ☆ **Chandra Laksha:** Laccadive Ordinary x Chawghat Orange Dwarf.
- ☆ **Laksha Ganga:** Laccadive Ordinary x Gangabondam.
- ☆ **Kera Ganga:** West Coast Tall x Gangabondam.
- ☆ **Anantha Ganga:** Andaman Ordinary X Gangabondam.

Other Varieties

Varieties released by TNAU, Coconut Research Centre – Veppankulam are VPM –3, VHC –1, VHC –2, VHC-3 and ALR-1.

Production of Coconut Seedlings

Since coconut cultivation involves substantial pre-bearing investment, greater emphasis must be given to the selection of the right type of planting material.

Quality seed nuts and seedlings are obtained through a series of selections made at various stages.

Collect seed nuts from February to May and harvest only fully matured nuts (11-12 months old).

Planting

Dig a pit of 1 m^3 for planting. Under a high water table area, planting is done on the surface or even on mounds. Planting distance is to be decided on the basis of whether the crop is grown in monoculture or in association with other crops. In India, spacing of 7.5 to 9 metres is recommended for pure crops of Ordinary Tall cultivars.

Manures and Fertilizers

There are various bases to recommend manures and fertilizers. In coconut, leaf and nut water are suitable for the diagnosis of nutrient deficiency through chemical analysis. The leaf sampling gave a more accurate analytical value than nut water and it is also easier. Leaf sampling technique is based on a critical nutrient level concept, which may be defined as the level of nutrients in the diagnostic plant parts below which the application of that nutrient in appropriate form and quantity is expected to increase yield. For bearing palm, it is suggested that the index leaf is derived by the formula n/2+1, whereas n is the total number of leaves present on the crown of the tree. In young palms, the chloride content in the seventh frond (leaf like structure in coconut palm) has been well correlated with growth parameters like height, development and number of leaves. Younger leaves are for Cu and Boron and older leaves are (14th leaf) for Fe and Mn status of the palm. In most of the cases 14th leaf, starting from fully open leaves, is most commonly used for chemical analysis. Crown choking in coconut is due to boron deficiency. Boron deficiency also causes barrenness of nuts and rotting of crown, bud and leaf malformation. Zinc deficiency causes dwarf and deformed growth of tender leaves called rosette or little leaves.

Table 23.1: Fertilizer Recommendation by Central Plantation Crop Research Institute

Name of fertilizers	*Dose (kg/year/palm)*	
	Ordinary tall under rainfed condition	*Hybrid tall under irrigated condition*
Ammonium sulphate or CAN	1.7	2.5
Urea	0.7	1.0
Single Super Phosphate	1.0	2.0
Muriate of Potash	1.2	2.0

The appropriate time for fertilizer application is when soil is not too dry or too wet. Generally, periods of heavy rains and long dry spells are avoided for fertilizer application.

Irrigation

This crop requires a large quantity of water for its normal growth and functioning. A single functioning root absorbs about 400 ml water per day, and it may have about 4000-7000 roots under favourable conditions. Coconut inflorescence carries about 16 female flowers at the time of opening of a spathe and about 2/3rd of these are lost through immature nut fall due to inadequate supply of moisture over a period of 6 months of nut development. In drip irrigation, 30-40 liters of water per plant per day is optimum for West Coast conditions.

In a study, it was observed that the application of irrigation at 10 day intervals achieved 53.5 per cent more yield as compared to the application of irrigation at 15 day intervals.

Inter and Mixed Cropping

Coconut plants are planted at 7.5 m x 7.5 m spacing. In this spacing, about 75 per cent of the area and half of the sunlight remain unused. To utilize space and sunlight, suitable intercrop can be grown after attaining the palm's height 5-6 m. The major suitable crops are banana, pineapple, groundnut, sweet potato, chillies and tapioca. In older plantations, the crops like pepper, cocoa, clove, cinnamon and nutmeg can be grown as mixed crops.

The coconut crop requires regular irrigation during the summer months in areas with uneven distribution of rainfall.

Problems in Coconut

1. **Shedding of a button and immature nut:** It is one of the most important factors which adversely affects the yield to a larger extent. The possible cause of this disorder may be pests and diseases. It has also been reported that the falling of buttons and immature nuts may be due to the fungus *Phytophthora* spp.

   The insects that cause considerable damage to female flowers are moths, beetles and bugs. The second factor is nutritional deficiency and it has been found that the application of nitrogenous fertilizers increases the fruit setting and ultimately yield.

2. **Unfavourable climatic conditions:** It is not only the dry condition that favours the shedding of nuts or buttons, but during the wettest time of year when the soil moisture is sufficient, the shedding of buttons and immature nuts is also there. It was observed that the problem was more serious in heavy soils than in lighter soils.

   Defective pollens and fertilization are also considered to be one of the causes of the shedding of immature nuts and buttons, but when artificial pollination is done, it did not result in the shedding of buttons.

3. **Production of barren nuts:** Some plants are producing barren nuts. These nuts are oblong in shape and have less amount of husk production

as compared to normal nuts. In barren nuts, generally, the embryo is absent, or if present, it is in varying stages of decay. There are various causes which leads to the production of barren nuts:

i) Defective fertilization results in the malformation of the embryo.

ii) Nutritional deficiency.

iii) Bearing more fruits may be one of the reasons.

Harvesting

Coconut yields 80-100 nuts/palm/year. Coconut usually matures in about 12-13 months after the opening of inflorescence. To get the maximum yield of copra and oil, the mature nut should be harvested, if they are harvested immature, the yield of copra and oil is considerably reduced depending upon the time of immature harvesting. The fully mature nut has 30-40 per cent coir. Coconut water contains 94.5 per cent water + vitamin C + vitamin B + sugar +fibre.

Post-harvest Management

- About 55 per cent of coconut production is consumed raw.
- After harvesting fruits, the coconuts are converted into copra and coconut oil.
- The husk of coconut is used in the manufacturing of coir mats, cushions and other products.
- Copra is manufactured in two forms. Milling copra is manufactured at commercial level and it is most popular in South India. It is used to extract oil whereas, copra is used to eat dry.
- In Kerala, 60-65 per cent of coconut production is converted into milling copra. Tapping of coconut (sap) is done from unopened inflorescence for toddy.

Toddy: It is a non-alcoholic beverage.

Kallu: Sour beverage made from fermented sap.

Insects

- **Rhinoceros beetle:** It damages unopened leaves and spathes by chewing off the internal tissues.

  Dispose of dead and decayed debris. Apply powdered neem cake along with an equal amount of sand in the axil of the top three leaves.
- **Eriophyid mite:** It damages the buttons during the early stage or soon after the pollination.

  Apply neem oil or garlic mixture @ 2 per cent or botanical pesticide azadirachtin 10000 ppm @ 0.004 % or 7.5 ml azadirachtin of 5 per cent formulation in the roots.

- **Red palm weevil:** It damages the plants during the early period of growth and results in the drying of young heart leaves.

  Keep the orchard free from dead and dried debris. Use the trap containing ferruginol (one trap per hectare) at about 2 m in height from ground level.
- **Leaf eating caterpillar:** Larvae feed on the lower surface of the leaf and decrease the photosynthetic efficiency of the plant.

  Cut and burn the fully infested leaves. Release 20 parasitoids (Bethylidae) per plant at 3rd larval stage. Apply malathion @ 0.05 per cent.
- **White grub:** It is 'C' shaped white coloured larvae with a black head and feed the underground portion of the plant.

  Deep plough during post-monsoon season. Apply chlorpyriphos 20EC @ 0.05 per cent.

Diseases

- **Basal stem rot (*Ganoderma spp.*):** Clean cultivation and isolate or uproot the diseased plant.

  Apply *Trichoderma* @ 5 g + neem cake 5 kg mixture per plant.
- **Stem bleeding:** Apply *Pseudomonas fluorescens* in the crown region.

  Apply copper oxychloride @ 0.3 per cent.
- **Bud rot:** Remove the infected portion and apply Bordeaux paste on the cut portion.

  Spray Bordeaux mixture @ 1.0 per cent.
- **Root wilt:** Apply a balanced dose of fertilizer and avoid water logging.

  Spray Bordeaux mixture @ 1.0 per cent or Mancozeb @ 0.3 per cent.
- **Grey leaf blight:** Remove and burn the dead, diseased and infected fallen leaves periodically.

  Spray copper oxychloride @ 0.25 per cent or Bordeaux mixture @ 1.0 per cent before the onset of rains.

Chapter 24

Arecanut/Betel Nut/Supari (*Areca catechu* L.)

Family: Arecaceae/Palmae; 2n = 32

Introduction

- ☆ Arecanut: Central Plantation Crop Research Institute (CPCRI), Regional Research Station, Vittal, Karnataka.
- ☆ Directorate of Arecanut and Spices Development, Calicut, Kerala.
- ☆ It is cultivated for kernel production, which are being used for chewing in its tender, ripe or processed form.
- ☆ It is consumed both as a raw and ripe nut (adaka or kacha tamul), dried ripe nut (chali supari), semi-mature, cut and processed (Bateldike or Kalipak).
- ☆ Stimulating agent, arecoline (0.1 %) is present in arecanut.
- ☆ Polyphenols and tannins are responsible for the astringent taste of nuts.
- ☆ Central Arecanut Marketing and Processing Co-operative Ltd. helps growers in Kerala and Karnataka to get remunerative prices.
- ☆ Arecanut flowers are sessile, and fruits are single-seeded berries.
- ☆ The edible endosperm is an economic part of arecanut.
- ☆ In India, it is grown in an area of 7.94 lakh ha with a production of 15.63 lakh tons. Karnataka (62 %), Kerala (15.6 %) and Assam (9 %) are the leading states in arecanut production. Arecanut is an essential ingredient of *gutka* and *pan masala*.

- It is commercially grown in India, Bangladesh and Sri Lanka. India ranks first in area (48 %) and production (56.6 %) and 7th in productivity in the world.

Origin

Native of Malayan Archipelago, Philippines, and other East Indian Islands.

Soil and Climate

The site selected for arecanut should have adequate irrigation facilities.

- The soil should be deep and well-drained without a high water table.
- It is highly susceptible to sun scorch, so the areca palm should be protected from exposure to the southwest sun.
- It grows well from sea level to an altitude of 1000 m above mean sea level in areas receiving abundant and well-distributed rainfall, *i.e.* 750 to 4500 mm annually or in irrigated conditions.
- Areca grows well within temperature ranging from 14 to 36°C and is adversely affected by temperature below 10°C and above 40°C.

Cultivars

Tall

- **Sumangala:** It is a tall type, partially droop variety. Ripe nuts are yellow to orange and oblong round in shape. Yield is 3.28 kg chali per palm per year.
- **Sreemangala:** It is tall and has a partially droop crown with a large internodal distance. Ripe nuts are yellow-coloured and round to oblong in shape. Yield is 3.18 kg chali per palm per year.
- **Swarnamangala:** It is a tall and heavy yielder variety. Nuts are bigger in size and the yield is 3.88 kg chali per palm per year.
- **Mohitnagar:** Nuts are placed loosely and uniformly. Yield is 3.67 kg chali per palm per year.
- **Kahikuchi Tall:** It is a high-yielding and regular bearer variety. Nuts are round and bold. Chali yield is 3.70 kg per palm per year.
- **Madhuramangala:** It is a heavy yielder, suitable for tender and processing purposes. It fetches a good price in the market due to the better appearance of the split, and the yield is 3.54 kg per palm (dry kernel) per year.
- **Nalberi:** It is a heavy yielder, regular bearer variety. Nuts are round shaped and yellow coloured. Yield is 4.15 kg dry kernel per plant per year.

Semi tall

- **Shantamangala:** High-yielding variety suitable for tender and ripe nuts. Yield is 3.96 kg dry kernel/ palm/ year.

Dwarf hybrids

VTLAH 1: Hirehalli Dwarf x Sumangala

VTLAH 2: Hirehalli Dwarf x Mohitnagar

Propagation

Propagated by seed having an average seed nut weight of 35 g.

Selection of Mother Palm and Raising of Nursery

- Mother palm should be superior, early bearer with a regular bearing habit, a large number of leaves on the crown, shorter inter-node length, high fruit set and high yield.
- The selection of seed nuts should be done after stabilization of yield, *i.e.* 4-5 years from first bearing. Select only fully ripe nuts (35 g) as seed nuts and discard under-developed and malformed nuts. The nuts should float vertically, with the calyx end pointing upward.
- Selected seed nuts are sown soon after harvest at 5 to 6 cm apart in beds of sand under partial shade, with their stalk ends pointing upwards. Cover the sand over the nuts and irrigate regularly.
- Germination starts 40 days after sowing and sprouts are ready for transplanting in secondary nursery when they are 3 months old.
- Newly germinated seedlings or sprouts are transplanted during the onset of monsoon at 30 cm x 30 cm spacing.
- Apply farmyard manure @ 5 t/ha in the secondary nursery.
- Provide proper shade using coconut/arecanut/banana leaves.
- 12-18 months old seedlings would be ready for transplanting.
- The nursery should be kept clean and regular irrigation should be applied during summer.

Planting

- Seedlings are planted in May-June or September-October at 2.7 m x 2.7 m spacing and 90 cm depth in well-drained soils while 60 cm in heavy soils.
- Before planting, pits of 90 cm x 90 cm x 90 cm are dug and refilled with topsoil mixed with cow dung and sand.
- Arecanut is a monoecious crop. The inflorescence is called spadix, emerges from the leaf axil.

Manures and Fertilizers

- Apply 100 g N, 40 g P_2O_5, 140 g K_2O and 12 kg green manure or compost or cattle manure per bearing palm per annum.

- ☆ Apply fertilizer in 2 split doses, *i.e.* September to October and February. Apply green manure or compost in a single dose during September-October.

Irrigation

Palm should be irrigated once in 4 to 7 days, depending upon soil and climatic conditions. Apply 175 litres of water per palm per irrigation.

Shading

- ☆ Areca palms are highly susceptible to sun scorching. Protect the seedlings against direct exposure to the sun. This may be done either by covering plants with areca or coconut leaves or by raising crops like banana between two rows of areca.
- ☆ Scorching occurs the October to January. Young palms should be protected from the south-west side by growing quick-growing shade plants.

Drainage

- ☆ Make 15-30 cm deeper channels than the depth of seedlings bed for drainage.

Intercropping/Multiple cropping

- ☆ Multiple cropping and intercropping in the arecanut garden provide an additional income. Grow shade-loving intercrop, which does not compete with arecanut for various resources and should have market feasibility, *e.g.* pepper, cocoa, banana, pineapple, cassava, ginger, sweet potato, turmeric and betel vine.

Harvesting

- ☆ Pre-bearing age ranges from 5 to 8 years. The colour of the fruit changes from green to yellow and red during ripening.
- ☆ Nets are harvested at tender and mature stage.
- ☆ Tender nuts are harvested from July to December, and ripe nuts are harvested from December to March or May to July.
- ☆ Three to four plucking of nuts are required during the whole season.
- ☆ The stage of harvesting depends on the type of product to be prepared.
- ☆ Kalipak (processed green nuts) is prepared from immature green nuts.
- ☆ Chali or kottapak is most popular for trading prepared from dried ripe nuts, where as pieces of dried nuts blended with flavour is called scented supari.
- ☆ Nuli is made from tender nuts.
- ☆ Palm yields more than 10 kg of ripe nuts per palm after the 10th year of planting.

Insects

- **Mites:** Suck sap from the lower surface of leaves and leaves become yellow and bronzing.

  Controlled by spray with dimethoate @ 0.05 per cent.
- **Spindle bug:** It causes injury on the petiole or lamina, which results in the appearance of slow-drying brown patches.

  Spray dimethoate @ 0.05 per cent.
- **Inflorescence caterpillar:** Feed on inflorescence.

  To control spray malathion @ 250 ml/100 litre of water.
- **Root grub:** It feeds on the root and results in the yellowing of leaves, stem and fall of nuts.

  First to loosen the soil upto 10 to 15 cm depth around the main stem of palm and drench it with 0.04 per cent chlorpyriphos during May and September-October.

Diseases

- **Fruit rot:** Water-soaked lesions appear on the nut surface.

  Spray Bordeaux mixture @1 per cent on bunches 3 times a year.
- **Bud rot:** The affected spindle appears yellow, changed to brown and after that the whole spindle rot.

  Remove affected parts and drench the crown with 1.0 per cent Bordeaux mixture.
- **Dieback of inflorescence:** Occurrence of brownish spots and spread these spots from top to bottom of inflorescence.

  Affected inflorescence should be removed and apply Mancozeb @3.0 g/litre as foliar application during flowering and repeat after 20 days.

Chapter 25

Cashew (*Anacardium occidentale* L.)

Family: Anacardiaceae; 2n = 42

Introduction

- *Cashew*: Directorate of Cashew Research, Puttur, Karnataka.
- Directorate of Cashewnut & Cocoa Development (DCCD), Cochin, Kerala.
- It is believed that cashewnut was introduced on Malabar Coast in India during the 16th century and dispersed to other countries in Asia.
- But in the 20th century, the cashewnut kernel gained importance in the international trade among edible nuts.
- Cashewnut is one of the important tropical crops, and it ranks second in international trade among nuts.
- India ranks 2nd in the world in the area and production of cashewnut.
- In the past two or three decades, cashewnut has gained status in horticultural crops.
- India is the largest producer and exporter of raw cashewnut.
- Inflorescence is intermediate panicle and is polygamomonoecious (trimonoecy). Its panicle consists of male, female and bisexual flowers.
- Fruit is a drupaceous nut. Cashew apple is a fleshy peduncle.
- The cashew apple, also called cashew fruit, is the fleshy part of the cashew fruit attached to the cashew nut. This pseudo fruit (or "false fruit") is a byproduct of the cashew nut industry.

- The area under cashew in India is 1.84 m ha with a production of 7.52 lakh MT.
- Export earnings from cashewnut are 2 per cent of total earnings from agricultural commodities.
- The top end of the cashew apple is attached to the stem that comes off the tree.
- The bottom end of the cashew apple attaches to the cashew nut, which is encased in a shell.
- The cashewnut kernels have high nutritive value.
- It is rich in protein, carbohydrates, unsaturated fats, minerals like Ca, P, Fe and vitamins A, B and E.
- Cashewnut protein is completed with all essential amino acids, and the kernel supplies 6000 calories energy/kg against 3600 by cereals, 1800 by meat, and 1650 by fresh fruits.
- As the nut fats are complete, very active, early digestive and non-fattening, so these can be taken by old and infants alike.
- Cashew apple is a rich source of vitamin C (250-300 mg/100 g).
- Fenny is an alcoholic beverage prepared from cashew apple. Cashew nut shell liquid (CNSL) is a byproduct and is used for making of resins, varnish, paints, insecticides *etc.*
- Kernels contain 47 per cent fat, 21 per cent protein, 22 per cent carbohydrates and 10 per cent vitamins. Kernels do not have cholesterol.
- About 82 per cent of fat contains unsaturated fatty acids.
- It is a good nerve tonic, a steady stimulant and controls diabetes.
- Major cashew nut-producing states in India are Maharashtra (25.8 %), Andhra Pradesh (16.4 %), Orissa, Karnataka, Tamil Nadu, Kerala, *etc.*
- As many as 20 species of cashew nut are known to exist in Central and South America. USA is the largest importer of cashew kernel.

Origin

- This crop is believed to have originated from Brazil and later on dispersed in countries like India, Brazil, Kenya, Madagascar, Zambia and Tanzania.

Soil and Climate

- Cashews can be cultivated on a wide range of soils, and it is often mentioned that it can survive on wastelands with low fertility levels.
- Being a very hardy crop, soil depth, slope, soil fertility and availability of water are not limiting factors, but water logging and excessive soil salinity are limiting factors.

- For getting a better yield, sandy loam soil without a hard pan should be preferred.
- This crop can be grown in a wide range of climates; however, it is grown in tropical regions. However, it is sensitive to frost and exposure to very low temperatures for longer periods.
- It requires a minimum rainfall of 50 cm/annum, the cashew is a sun-loving tree and does not tolerate excessive moisture. It is very sensitive to water logging.
- Very high temperature, 39 to 42°C cause fruit to drop during the marble stage of fruit development.
- This crop can be grown up to pH 8, higher pH soils are not suitable for commercial cultivation of this crop.

Cultivars

De Candole (1825) subdivided *Anacardium occidentale* into two groups on the basis of the ratio of nuts to peduncle, *i.e.* American and Indian. Some of the important cultivars released in India are described as follows:

- **BPP-1:** It is a hybrid nut. Nuts are medium-sized and have a weight of 5 g. The yield potential is 17 kg/tree.
- **BPP-2:** This hybrid is similar to BPP-1, but the yield potential is 19 kg per tree per year at the age of 25 years. The weight of nuts is about 4 g.
- **BPP-3:** It is a selection from germplasm Sainha Chelam in Vishakhapatnam, with an average yield of 16 kg/tree and net weight of 6 g.
- The other cultivars are BPP 4, BPP 5, BPP 6, BPP 8, BPP 9, Vengurla 1, 2, 3, 4 and 5 from West Bengal. Selection from Tamil Nadu are BRI-1 and 2, Ullal 1 and 2.

Propagation Methods

- Cashewnut can be propagated both sexually and asexually.
- For sexual propagation, the seed should be collected from a compact, intensive branching, high proportion of bisexual flowers and cluster bearing habit.
- The vegetative propagation is done using softwood grafting and epicotyl grafting. Soft wood grafting is the best.
- The dwarfing rootstock of cashews is *Anacardium pumilum.*

Planting

Transplanting from the nursery to the field should be done by the onset of south-west monsoon at a spacing of 7 m x 7 m or 8 m x 8 m (90 or 72 plants/acre).

Manures and Fertilizers

In India, generally manuring is not done by the farmers, but regular fertilizer should be applied to ensure early and high yield.

Table 25.1: Application of Fertilizers in Cashewnut

Age of Plant	*May- June (g/tree)*			*September- October (g/tree)*		
	N	*P*	*K*	*N*	*P*	*K*
1st year	60	40	—	60	40	—
2nd year	100	40	30	100	40	30
3rd year	200	60	60	200	60	60
4th year	250	65	65	250	65	65

These fertilizers should be applied 25 cm away from the main trunk and mixed with the top 30 cm of soil.

Irrigation

- In India, most of the cashews are grown under rainfed conditions.
- Life-saving irrigation during summer months at fortnightly intervals, *i.e.* January-March, about 200 litre water per plant improves fruit set, fruit retention and fruit per nut yield.

Training and Pruning

All sprouts up to 2 m from the ground level should be removed. Removal of dead, diseased and criss-cross branches during July minimizes the loss in yield and decreases the incidence of dieback disease.

Top Working

- Most cashew plants are of seedling origin, so their yield and quality are low.
- To improve yield, 20-25 years old seedling plants are headed back to the 0.5 m above the ground level in the month of December- January. Apply Bordeaux paste over the cut portion of the trunk (stump) to avoid pathogen and insect pests (borer).
- Profuse sprouting will occur on the stump below the cut portion, retaining 10-15 healthy shoots, maintaining proper distance.
- Retain 7-8 well-sprouted healthy shoots and graft on these retained shoots at 40-50 days after sprouting or at the soft wood stage. These propagated plant starts yielding about 4 kg of nuts from the next year of rejuvenation/top working and gradually yield increases to 8 kg from the 4th year of top working.

Weeding

- Cleaning the canopy area manually within a 2 m radius of the trunk. Weeding is done at least twice a year.
- After emergence, weeds can be controlled with the application of glyphosate @ 6 to 7 ml per litre of water or 0.8 kg a.i./ha during June – July for effective weed control.

Harvesting

- Cashew tree starts yielding from 4 to 5 years of planting and reaches a peak about 10 years after planting. The productive life of plant remains about 40 -50 years.
- Flowering occurs from November to February.
- The harvesting time of fruit is February to May and peak harvesting time is March-April.
- The ripened fruits fall down and the fallen nuts are to be collected. The yield per plant starts from 1 kg in the 3rd to 4th year to 10 kg from 8-10 year old plants, which have developed a full-sized canopy. However, it depends upon management practices as well as the age of the plant.

Insects

- **Tea mosquitoes bug:** Nymphs and adults suck the cell sap from leaves, shoots, panicles and immature nuts. Its feeding causes the exudation of gummy substances.

  Spray lambda cyhalothrin (0.6 ml/litre), thiametoxam (0.2 g/litre) and acetamiprid (0.5 g/litre).
- **Stem and root borer:** Exudates sticky substances from the infected area and the leaves become yellowish and result in premature leaf fall.

  Drench the plants with Fipronil 2 ml/litre.
- **Leaf miner:** Miner injures both leaves and shoots. It feeds below the epidermal layer and causes curling and browning of leaves.

  Spray azadirachtin or Spinosad or quinolphos (1.5 ml/litre) or profenophos (1.5 ml/litre) or lambda-cyhalothrin (0.6 ml/litre) for effective management of leaf miners.

Diseases

- **Damping off of seedlings:** Water-soaked spots appear on the shoots and newly emerged plants near the ground level and the seedlings die.

  Soil drenching with copper oxychloride @ 3 g /litre or Bordeaux mixture 1.0 per cent or mancozeb @ 3 g / litre is effective in controlling the seedling infection.

- **Inflorescence blight:** Small-sized reddish-brown water-soaked spots appear and dry up of inflorescence.

  Spray copper oxychloride @ 0.2 per cent.
- **Dieback or pink disease:** White or pink-coloured fungus growth can be seen on the affected area. The penetration of fungus in deeper tissues causes drying of shoots.

  Spray wettable sulphur @ 0.2 per cent or carbedazim @ 1 ml/ litre of water.
- **Anthracnose:** Affected area shows water-soaked spots and red coloured lesions. Shoot dies when lesions enlarge.

  Spray Bordeaux mixture @ 1 per cent or mancozeb 0.2 per cent or copper oxychloride 0.3 per cent.

Chapter 26

Tea (*Camellia* spp.)

Family: Theaceae; 2n = 30

Introduction

Tea plants are classified into three different species. They are:

1. Assam tea (*Camellia sinensis* var. *assamica*).
2. China tea (*Camellia sinensis*).
3. Cambod tea (*Camellia assamica* sub sp. *lasiocalyx*).

Uses

1. **Beverages**: Tea contains alkaloids called thein (2 to 5 %), which is a caffeinaceous alkaloid that relieves body fatigue. However, excessive tea drinking is harmful to our digestive systems.
2. **Polyphenols:** Polyphenols reduce blood cholesterol and cure hepatitis, hypertension, stone formation *etc* (Heart attack is due to cholesterol + triglycerides which are saturated fatty acids).
3. **Medicinal properties:** Black tea (a fully fermented tea) is anti-ulceric and anti-carcinogenic in nature.

Area and Production

- ☆ In India, 80 per cent of the area is above 50 ha in size and most of it is in the corporate sector (United body).
- ☆ North India = Covers about 80 per cent of the total area of India.
- ☆ South India = Covers about 20 per cent of the total area of India.
- ☆ Tea is also called the Queen of Beverage crops.

  Distribution in India:

- ☆ Assam = 53 per cent.
- ☆ West Bengal = 24 per cent.
- ☆ Tamil Nadu = 13 per cent.
- ☆ Kerala = 8 per cent.
- ☆ Other states like Karnataka, Tripura, Himachal Pradesh, Uttar Pradesh, Bihar, Sikkim and Manipur = 2 per cent.
- ☆ Although Darjeeling produces the finest quality tea, however, its productivity is low. The average yield in Darjeeling (635 kg/ha) is reported to be less than half of the all India average.

Research Stations Board working on tea cultivation in India

1. UPASI: United Planters Association of Southern India, UPASI Tea Research Institute, Nirar Dam, Valparai, Coimbatore, Tamil Nadu.
2. TES: Tea Experiment Station, Tocklai, Jorhat, Assam.
3. Tea Board: Calcutta.

Table 26.1: Different Features of Assam Tea and China Tea

Sr. No.	*Features*	*Assam tea (Camellia sinensis var. assamica)*	*China tea (Camellia sinensis)*
1	Stature	Tree	Shrub/bush
2	Branches and growth rate	Few, robust branches and quick growing.	Abundant branches and slow growing.
3	Leaves	Large (15 to 20 cm long), glossy and less serrated.	Small (4 to 10 cm long), leathery and more serrated.
4	Quality	Medium (better strength).	Good (Better flavour).
5	Yield	High yield.	Low yield.
6	Susceptible to drought and frost	Susceptible.	Hardy and tolerant.
7	Distribution	Countries near the equator *viz.*, India, Sri Lanka, S.E. Asia, Central Africa *etc.*	Cold countries and high-altitude areas like Japan, N-Iran, S. Russia and China *etc.*
8	Economic life	Less (40 years).	More (over 100 years).

Origin

Chinese people are the first users of tea.

The region from South East China (Sichuan-Yunnan) to Assam (India) has been reported as the centre of origin of the tea plant.

Soil and Climate

- ☆ The tea performs better in well-drained soils with a good amount of organic matter and acidic soil reactions with pH 4.5 to 5.5.

- ☆ Being a subtropical crop, it performs well at a height of 1000 - 2500 m from the sea level.
- ☆ Tea prefers moist, warm and winter (not too cool) with an optimum temperature of 20-27°C and well-distributed rainfall from 125 to 750 cm.
- ☆ The day length of 10.5 hours is critical for the vegetative growth of tea.

Calcifuge Nature of Tea

- ☆ Tea is not performing satisfactorily with high calcium soils. Tea is a calcifuge (does not tolerate alkaline soils) and is an aluminium accumulator. It performs best under acidic soil conditions.
- ☆ The soils with high amounts of iron and manganese are good for tea cultivation.

Nursery

- ☆ The nursery soil should be well drained and deep loam in nature with pH 4.5 to 4.8.
- ☆ The soil and sand used in the preparation of the rooting medium should be tested for pH and nematode infestation.

Manuring of Nursery

After removing the tent, the cuttings are sorted and staked. Apply 30 g of the following mixture in 10 litres of water for 4 m^2 of nursery area at fortnightly intervals.

Table 26.2: Fertilizer Composition

Fertilizers	*Quantity*
Ammonium phosphate (20:20)	35 parts by weight
Potassium sulphate	15 parts by weight
(or)	
Muriate of potash	12 parts by weight
Magnesium sulphate	15 parts by weight
Zinc sulphate	3 parts by weight
Total	80 parts by weight

Propagation

i. Seed/Sexual

- ☆ Tea is generally propagated by seeds. It is the easiest and most convenient method of propagation.

- Now-a-days, some high-yielding varieties are propagated vegetatively because tea seeds are not true to type.
- The seedling plants are heterogeneous in nature.
- Along with good yield, the plant should have good manufacturing quality and are resistant to diseases also.

ii. Vegetative Propagation

The vegetative methods for the propagation of tea are as under:

1. Cutting (this is a popular method for vegetative propagation).
2. Grafting
3. Budding
4. Layering

However, budding, grafting and layering are laborious methods of propagation.

i) Cuttings

The commercial method of vegetative propagation is single node semi-hard wood cuttings. Cutting must be taken from the mother plant that has desirable characteristics like high yield, good quality and quick recovery.

ii) Grafting

Drought susceptible high yielding clones can be used on the drought-hardy rootstocks. Drought susceptible high yielding scion clones are UPASI -3, UPASI – 8 and UPASI -17. Drought hardy rootstock clones are UPASI-2, UPASI-6, UPASI-9, UPASI-10, UPASI- 24, ATK-1, SA-6 and TRI 2025 (UPASI: United Planters Association of Southern India).

Features of Mother Bush

1. Stem should be thick, uniform, compact and well distributed.
2. Bushes should have an early filling tendency and lateral spreading habit.
3. The mother bush should have upright growth of the main stem and branches (primary, secondary and tertiary) after pruning.
4. More plucking points or more plucking density should be there.
5. There should be pubescence in sprouts or new flushes/leaves.
6. The plant should be tolerant to drought.
7. The plant should be resistant to blister blight and free from insect pests.
8. The mother bush should be a high yielder per unit area.
9. Bushes should not have a flowering tendency and too many bhanji buds (buds on the shoot remain dormant or don't develop into active buds).

Hardening of Cuttings

Cuttings are hardened by decreasing the shade gradually or exposing the cuttings to sunlight or an open environment. Initially, the cuttings are exposed for a short period of exposure to the sun and gradually increase the time of exposure to the sun.

Mother Bush Treatment

High-yielding, vigorous and healthy mother bush should be selected. Apply 40 g of young tea mixture + 60:90 N:K mixture up to 5 years to each selected bush. The mixture of the following chemicals or fertilizers should be applied to the mother bush before the collection of cuttings.

- 0.5 per cent $AlSO_4$ + 1 per cent $MgSO_4$ (before 3 weeks).
- 2 per cent $ZnSO_4$ (before 2 weeks).
- 1 per cent urea (before 1 week).

Planting

The economic life of a tea bush is over 40 years. Following points should be kept in mind while selecting a site:

1. **Sloppy land or hilly terrains or jungles:** To conserve soil moisture and protect erosion, adequate soil conservation measures should be taken care.
2. **Soil**: Well-drained, deep with acidic reactions are suitable for planting.
3. **Temporary shade:** Provide shade to the young developing plants.

Time of Planting

North India: In Assam conditions, the winter season has low rainfall, but a sufficient amount of mist is there for planting, *i.e.* during October– November.

South India: In South Indian conditions, the summer season (May – June), *i.e.* the rainy season is suitable for planting.

Planting Operation

Pit size: Hole or trenches of 45 cm x 45 cm x 60 cm (depth).

Or pits of 30 cm x 30 cm x 45 cm (depth).

If the land is not good.

45 cm x 45 cm x 45 cm (depth).

60 cm x 60 cm x 60 cm (depth).

Spacing

Spacing between plants varies from region to region. The optimum plant population is 1400 to 1800 per hectare. More density adversely affects the plant growth after 4-5 years of planting.

1.5 m x 1.5 m or 1.5 m x 0.75 m.

1.2 m x 0.75 m or 1.2 m x 0.60 m.

Methods of Planting

- **Single hedge system**: Spacing 1.20 m x 0.75 m.
- **Double hedge system:** Spacing 1.35 m x 0.75 m x 0.75 m.

Planting Season

May - June or September – October.

Make open sleeves between the rows without disturbing the roots. The pit around the stem should be gently pressed after planting, or the air pockets in the root zone should be removed for proper contact of roots with soil.

Manuring

Apply manures two months after planting. Add 80-100 kg/ha phosphorous as rock phosphate by placing it at 15 to 25 cm depth once a year up to the first pruning, and after that, apply it once in two years. N: K should be applied in 2: 3 ratios for the first 3 years and thereafter in 1: 1 ratio.

Table 26.3: Application of Fertilizers

Year of application/ age of plant	*Total weight kg/ha/annum*		*Number of applications*
	N	*K*	
1st year	180	270	5
2nd year	240	360	6
3rd year	300	450	6
4th year onward	300	300	6

Fertilizers should be applied under the canopy of the plant or in the drip circle. It should not come in contact with the main stem or apply away from the main stem before the onset of monsoon.

Irrigation

Regular and subsoil irrigation should be given during the summer months for proper growth of seedlings.

After Cultivation

Weed control: For perennial weeds (Forbicot weeds), apply glyphosate 1.75 litre + Kaolin 2 litre + 2 kg of wetting agent in 450 litre.

For dicot weeds, apply gramoxone 500 ml in 200 litre water.

Training Young Tea

- **Centering:** Centering is done 3-5 months after planting for more lateral sprouts. Head back to the main stem, leaving 8-10 mature leaves.
- **Tipping:** It is done 60 cm above the ground level or 35 cm above the first tipping.

Pruning

Pruning is done to remove the dead and diseased portion of the plant. It is also helpful to maintain proper plant or bush height.

$$\text{Area to be pruned every year} = \frac{\text{Total extent of the garden}}{\text{Pruning cycle}}$$

Pruning interval = (Elevation in feet/1000) + 1.

The appropriate pruning time is April - May or August - September.

Types of Pruning

- **Rejuvenation pruning:** It is being done to rejuvenate the plant. In this pruning, cut the main stem of the bush near the ground level or less than 30 cm height from the ground level.
- **Hard pruning:** This is done to give the bush proper shape and spread. In this pruning, cut 30-45 cm branches.
- **Medium pruning:** This type of pruning is done to stimulate new growth and increase foliage up to 60 cm bush height. It also checks the inconvenient growth of bush.
- **Light pruning:** It is done about an inch or less on the basis of previous pruning to a bush height less than 65 cm.
- **Skiffing:** It is done to maintain a uniform surface of the bushes (light pruning). Only the top portion, about 5-8 cm is removed. It is done to a bush height of more than 65 cm.

Management of Shade in Tea Plantation

Direct exposure of tea bushes to sunlight adversely affects growth.

Plants which have semi-erect leaf habits perform better under open conditions or do not require shade because leaves are not over-heated and bushes allow sunlight in the interior portion of the canopy. So, bushes with semi-erect leaves have higher yields than bushes with horizontal leaves.

However, broad-leaf bushes require shade and perform better under shade conditions.

1. **Protection:** Protect leaves from direct sunlight and UV radiation.
2. **Fertility improvement:** Shade tree adds about 8-10 t organic matter/ha/year and increase fertility of the soil.

Note: (a) The fertility level of the soil increases three times at temperature 32°C compared to temperature 16°C. (b) 22 t per hectare or 2.2 kg/m^2 organic matter is required to increase 1 per cent organic matter in the soil.

3. **Breaking up of subsoil:** Penetration of shade plants' root helps to break the hard pan of subsoil.
4. **Temperature effect:** Shade trees regulate temperature and humidity in the crop.

   Summer: Shade trees decrease temperature by 5 to 7°C in summer during the day time and increase temperature by 5 to 7°C during the night as compared to open areas.
5. **Windbreaks:** Shade trees act as windbreaks and decrease wind speed.
6. **Erosion control:** Litter fall adds organic matter and roots of the plant reduce soil erosion.
7. **Additional income**: Increase the yield of crops and provide timber and fuel.
8. **Decrease in pest incidence:** Less mite attack was observed in the shade. *e.g.* 24 mites per leaf were seen under shade. However, 61 mites per leaf were seen under open space.
9. **UV radiation:** Shade trees reduce sunburn and protect from UV rays. About 30-40 per cent of the rays are trapped by trees during full sun, *i.e.* mid day and minimize the effect of harmful radiations (UV).
10. **Increased soil fertility:** Litter fall increases fertility level by adding 8-10 t of organic matter/ha/year.
11. **Regulate humidity and decrease evaporation losses:** Tree shade regulates humidity and decreases temperature and evapotranspiration losses.

Annual Lopping

Cut the erect type branches on the laterals of shade trees before monsoon season.

Crop Duration and Harvest

The bushes are ready for plucking 3 years after planting, in plucking two leaves along with a bud, which is called "Two leaves and a bud", while in less plucking, plucking is done less than this. In South India, plucking is done throughout the year, while during March - May, it is done at weekly intervals and for a remaining period about fortnightly intervals.

- **Rush period**: The interval in the plucking period is less, *i.e.* 7 to 10 days.
- **Lean period**: During the lean period, plucking is done at 10 to 15 days intervals.

Yield

The yield of green leaves is 10 t/ha.

Grade Specification

- ☆ The size of the tea leaves, orange pekoe is referred to as the highest-grade tea. Whole, large tea is also graded as higher.
- ☆ Method of production: CTC process has strong strength (crush, tear and curl) and orthodox light-strength tea.

Insects

- ☆ **Scales:** It causes yellowing of leaves and stunted plant growth. Plants may die when infestation is more.

  Scales can be controlled by a spray of quinalphos 25 EC @ 2 ml/litre or chlorpyriphos 20 EC @ 2 ml/litre.
- ☆ **Sahyadrassus/Phassus borer:** It makes holes in the stem and feeds at night. Powdered silk hung near the hole.

  Insert a thick wire in the hole of the borer and kill the larvae.
- ☆ **Thrips:** It feed chlorophyll of leaves and shows a silvery surface on the leaf. Spray any one of the following insecticides.
  - Azadirachtin 1.0 per cent EC (neem based) @ 2.0 ml/litre.
  - Azadirachtin 5 per cent neem extract concentrate @ 5.0 ml/10 litre.
  - Profenophos 50EC @ 2.0 ml/litre.
  - Ethion 50 EC @ 5.0 ml/10 litre.
  - Quinalphos 25 EC @ 7.5 ml/10 litre.
- ☆ **Aphids:** It feed on tender leaves or newly pruned shoots. Leaf curl upward and plants become stunted.

  Spray phosalone 35 per cent EC @ 2.0 ml/litre.
- ☆ **Red spider mite:** Leaf becomes brownish and dry or falls during severity.
- ☆ **Pink mite:** Infested leaves turn pale and curve upward. In severe infestation, leaves become leathery brown.
- ☆ **Scarlet Mite:** It causes brown discolouration of leaves and scarlet can be seen with the naked eyes.

  Azadirachtin 1.0 per cent EC (neem based) @ 2.0 ml/litre.

  Azadirachtin 5 per cent neem extract concentrate @ 5.0 ml/10 litre.

  Fenazaquin 10EC @ 1.6 ml/litre.

  Ethion 50EC @ 5.0 ml/10 litre.

  Fenpyroximate 5 EC @ 1.2 ml/litre.

Hexythiazox 5.45 EC @ 1.2 ml/litre.

Fluphenzine 20SC @ 5.0 ml/10 litre.

Phosalone 35 EC @ 1.0 ml/litre.

Propargite 57 EC @ 2.0 ml/litre.

Profenophos 50EC @ 2.0 ml/litre.

Spiromesifen 22.9 SC @ 1.0 ml/litre.

☆ **Tea mosquito bug:** It sucks the sap from leaves by puncturing them. Water-soaked spots appear from the puncturing area.

Monitor the tea garden regularly for tea mosquito bugs.

Avoid the alternate hosts for tea mosquito bugs like neem, guava and cashew.

At the initial phase of infestation or during less infestation, spray any one of the following:

- Imidacloprid @ 0.6 ml/litre.
- Profenophos @ 2 ml/litre.
- Thiamethoxam @ 0.6 g/litre.

Diseases

☆ **Blister blight:** Very small spots seen on young leaves. Later on, spots become transparent. The lower surface of the leaf develops blister symptoms.

Spray 210 g copper oxychloride and nickel chloride per ha at 5-day intervals from June to September and 11-day intervals in October and November. Spray hexaconazole 200 ml + copper oxychloride 210 g at 5-day intervals/ha.

Chapter 27

Coffee (*Coffea arabica* L., Arabian Coffee; *Coffea canephora* Pierre ex. A. Froehner, Robusta Coffee)

Family: Rubiaceae; 2n = 22

Introduction

- Coffee: Central Coffee Research Institute, Balehonnur, Karnataka.
- It was introduced in India in 1600 AD by a Muslim pilgrim, Baba Budhan, in Karnataka.
- The average productivity is 947 kg of clean coffee per hectare.
- The fruit of coffee is drupe. Normal berry (2 seeded), pea berry (single seeded), triangular berry (three ovules in trilocular ovary) and elephant bean has different forms of the seed berry in the coffee.
- Non-fertilization or abortion of embryo after fertilization is the main cause of peaberry formation in coffee.
- Coffee is a single-seeded fruit called berry.
- Pollination in coffee is by wind and bees.
- Taferikela is a popular clone of coffee used as a male parent in a breeding programme. Coffee contains niacin, which is useful to cure skin diseases.
- It is being grown in about 2.43 lakh ha with production of 1.75 lakh tones covering Karnataka (56 %), Kerala (25 %), Tamil Nadu (9 %) and the remaining 10 per cent in non-traditional states like Andhra Pradesh, Odisha and North-Eastern states. Arabica shares 49.4 per cent, while

Robusta 50.59 per cent and 60 per cent cropped area. The export earnings are Rs. 300 crore per annum.

- ☆ In world, it is the second largest commodity in trade after petroleum products.

Three main species of coffee

1. *Coffea arabica* (Arabica coffee), it accounts for 75-80 per cent of the world's production and it is susceptible to pests and diseases.
2. *Coffea canephora* known as Robusta coffee, produces an inferior taste beverage with higher caffeine content. It is tolerant to pests and diseases.
3. *C. liberica* (tree coffee) beans are used for flavour.

Table 27.1: Difference between Arabica and Robusta Coffee

Sr.No.	*Character*	*Arabian (C. arabica)*	*Robusta (C. canephora)*
1	Ploidy	Tetraploid (2n=44).	Diploid (2n=22).
2	Adaptability	Higher elevation.	Lower elevation.
3	Plant status	Small tree, a shrub or a bush under training.	It is bigger tree than Arabica.
4	Leaves	Dark green.	Pale green.
5	Flowers	1-4 flower/inflorescence.	5-6 flower/inflorescence.
6	Berries	10-20 per node, oblong to round in shape.	-

Origin

Native of forests of Ethiopia and Central Africa.

Soil and Climate

It is grown partially in the shade of trees, which provides natural habitat. Tree also provide a proper microclimate for tropical rain forests. Scuffling (soil stirring) is practised in coffee.

Varieties

- ☆ **Kent:** It is the earliest variety of Arabica selected by an English planter in 1920.
- ☆ **S 795:** The most popular Arabica selection released during the 1940's had high yield, bold beans of superior quality and relatively tolerance to leaf rust. It is a selection from Kent.
- ☆ **Cauvery:** Popularly known as Catimor. It is high-yielding and resistant to pests and diseases. It is cross of a Caturra and Hybrido-de-Timor.
- ☆ **Selection 9:** Selection 9 is a derivative of a cross between an Ethiopian Arabica collection, "Tafarikela" and Hybrido-de-Timor. This variety has won the Fine Cup Award for the best Arabica at the Flavour of India-Cupping Competition 2002, organized by the Coffee Board of India.

- ☆ **Typica:** This is the base variety from which many coffee varieties have been developed. Typica coffee has very low production but excellent cup quality.
- ☆ **Bourbon:** Its plant produces 20-30 per cent more coffee than Typica but has a smaller harvest than most of the coffee varieties. The cup quality is excellent and similar to Typica.
- ☆ **Caturra:** It is the mutant of Bourbon coffee discovered in Brazil. It has high production and good quality, but requires extensive care.
- ☆ **Catuai:** It is a high-yielding cross of Mundo Novo and Caturra. Suitable for strong wind or high rain areas.
- ☆ **Pache Comum:** Mutant of Typica Coffee.
- ☆ **Pache Colis:** Coffee fruits are very large. It is resistant to phoma.
- ☆ **Catimor:** It is a cross of Timor coffee (resistance to rust) and Caturra coffee. It was developed in 1959 in Portugal. It is early maturing and very high-yielding.
- ☆ **Mundo Novo:** It is a natural hybrid between Typica and Bourbon coffee. It was first found in Brazil.
- ☆ **Maragogype:** It is a mutant of Typica and was discovered in Brazil.
- ☆ **Amarello:** It produces yellow fruit and is not widely planted.
- ☆ **Blue Mountain:** It is famous for its resistance to coffee berry disease and ability to thrive in high altitudes. It was first grown in Jamaica.

Other varieties: Sln.5A, Sln.5B, Sln.6, Sln.12, S.274, S.270.

Propagation

- ☆ In India, the seeds of coffee are being used for propagation. This is the easiest and simplest method of propagation. By this method, a large number of plants can be produced in a short period. In Arbica coffee, a reasonable level of variability can be achieved by selfing because this variety is self-compatible in nature. However, cross-pollinated cultivars (Robusta) show variability.
- ☆ True-to-type Robusta plants are propagated by rooted cuttings.
- ☆ For the multiplication of Arabica and Robusta, fully matured normal fruits are harvested at $3/4^{th}$ of the ripening stage from healthy and certified seed blocks during November-December and January-February, respectively.
- ☆ First to discard the floating fruits. Select healthy fruits and separate the seeds from the pulp using a sieve. Discard and remove the infected/defective and unhealthy seeds.
- ☆ The beans are mixed with wood ash @ 0.75 kg/kg of seed and dried under the shade, stirring from time to time to facilitate uniform drying. Seeds are treated with carbendazim @ 1.0 g or Vitavax @ 0.66 g/kg seeds.

Nursery

Seedbeds of 6 m x 1 m raised 15 cm from ground level are prepared under overhead pundal, using a mixture of well-sieved forest soil, FYM and sand. Seeds are usually sown during December- January at 2.5 to 3.0 cm depth in a regular row and covered with a thin layer of finely sieved soil.

Planting

In April, pits of 45 cm x 45 cm x 45 cm are dug as per the spacing given below and kept open up to monsoon.

- Tall Arabica like S 795, S 288: 2.1 m x 2.1 m.
- Semi-dwarf like Cauvery: 1.8 m x 1.8 m.
- Dwarf like S-7 (San Raman): 1.5 m x 1.5 m.
- Hybrid like Congensis X Robusta: 2.5 m x 2.5 m.
- Robusta Selections like S 274 and BR Series: 3 m x 3 m.

Planting is done after the end of heavy monsoon in August-September. At the time of planting, add 50 g of rock phosphate to each pit to encourage root growth.

Manures and Fertilizers

Add FYM 12.5 q/ha per year. The NPK should be applied at the proper time and doses as given below:

Table 27.2: Application of Fertilizers

Arabica	*Pre-blossom (March)*	*Post blossom, pre-monsoon (May)*	*Mid monsoon (August)*	*Post monsoon (October)*	*Total*
1st year	15:10:15	15:10:15	—	15:10:15	45:30:45
2nd and 3rd year	20:15:20	20:15:20	—	20:15:20	60:45:60
4th year	30:20:30	30:20:30	—	30:20:30	80:60:80
Bearing coffee 5 years and above: for less than 1 t/ha crop	40:30:40	40:30:40	20:0:0	40:30:40	140:90:120
For 1 t/ha and above	40:30:40	40:30:40	40:30:40	40:30:40	160:120:160
Robusta					
For less than 1t/ha crop	40:30:40	—	—	40:30:40	80:60:80
For 1 t/ha and above	40:30:40	40:30:40	—	40:30:40	120:90:120

Training and Pruning

Topping: Removal of the terminal portion of the main stem from the prescribed height.

(1) Single stem system (2) Multiple stem system

- Single stem training is followed in India for growing coffee under shade. Young plants are trained to provide the proper shape of bushes and to improve the efficiency of spray and harvest. The pruning system followed in coffee is a single stem open centre system. Plant height is restricted by topping (capping), which is a cultural practice in coffee.
- Tall Arabica varieties are topped at two stages (2-tier system), while the dwarf Arabica and Robusta are capped at a single level (single-tier system). Pruning is practised in June- July to August-September.

Topping Height

- **Tall Arabica:** 1st topping at 0.9 m; 2nd topping at 1.5 to 1.7 m (after harvesting 4-5 crops, when spread of lower canopy completed).
- **Dwarf Arabica:** Single topping at 1.0 to 1.7 m depending upon soil fertility and wind proneness.
- **Robusta:** Single topping at 1.5 to 1.7 m.

Pruning

- It is an evergreen plant, so to maintain a balance between vegetative growth and cropping wood, annual pruning is required.
- In Arabica, pruning should be done once a year after harvesting.
- The plants affected with die back are pruned after rainfall, so the plant may regenerate its new vegetative growth.

Criteria for pruning

- The primary pruning cannot be regenerated easily, so never cut these branches.
- If the primary branch is exhausted, then cut it, leaving some nodes for regeneration.
- The removal of secondary and tertiary branches can produce 2 to 3 crops and also encourage new branches.
- Remove upwards, downwards, inward and criss cross growing branches.
- Remove diseased and damaged suckers.
- Remove lean and lanky branches
- Robusta is a self-pruning plant. Remove suckers, diseased, insect-infested shoots 3-4 times a year. Under irrigated conditions, light annual pruning is required after harvesting of the crop.

Handling, Centering and De-suckering

Handling: It is thinning out of the young flesh, or sprouts arise after main pruning. The repetition of handling depends on the characteristics of the variety, it can be done once or twice, *i.e.* during June-July (1st handling) and during September (2nd handling) if required. Keep one healthy sprout on each side of the node and remove the remaining fleshes. Each primary branch can have 4-6 secondaries.

Centering: Remove all new arising shoots within a 15 cm radius from the primary branch (Centering) for proper aeration to the main stem.

Desuckering: Removal of suckers, these should be removed 3-4 times a year.

Planting of Shade Trees

- Plant temporary shade trees like dapdap (*Erythrina lithosperma*) initially or provide optimum shade to young coffee plants at a spacing of 5 to 6.5 m. In open spaces shade is provided by planting evergreen shade trees at a distance of 10-13 m. Arrange the shade in two tiers, 1st tier of low height for temporary shade and 2nd tier as a top canopy for permanent shade.
- The optimum shade requirement for Arabica is 50 per cent and for Robusta 30 per cent.
- Ideal shade brought down the incidence of pests like borer, green scale and diseases like leaf rust and black rot in Arabica.
- Just before monsoon, the *Erythrina* is topped for proper aeration and penetration of light. Regulate the canopy of permanent shade trees once in every 2 to 3 years to provide optimum shade.

Intercropping

The non-exhaust crops like elephant foot yam, ginger, banana, pineapple, papaya, *etc.* can be grown till the start of bearing.

Soil Management

It includes the conservation of soil and water described under following heads:

a) **Digging**: Digging of the field about 35-45 cm after the end of monsoon. In digging turn soil, remove stumps, burry weeds and remove vegetable debris.

b) **Scuffling or Soil stirring:** Scuffling or soil stirring is done to remove weeds and conserve moisture at the beginning of the dry period.

c) **Mulching:** Mulching is done to add fertility, suppress weed growth, conserve moisture, maintain soil temperature and control soil erosion.

d) **Trenching:** Trenching is done about 50 cm wide and 25 cm deep between the rows during August- September towards the contour when soil is easy to work.

e) **Weed control:** Remove weeds three to four times a year.

Remove weeds during the monsoon and after completion of the monsoon. Spray Gramoxone @ 1.25 litres per hectare in 450 litres of water.

f) **Irrigation:** Irrigate the crop by sprinkler irrigation for good blossoming during failure of rainfall.

g) **Soil acidity and liming:** The heavy rainfall caused the leaching of calcium and magnesium, and increased soil acidity. At the same time, the application of acid-forming fertilizers also increases acidity. Apply lime and dolomite lime to reduce the acidity.

Harvesting

1. Stripping method.
2. The method uses a comb to brush the trees.
3. Mechanical method.
4. Hand-picking after every 8-10 days.

- ☆ The dry season is best for harvesting coffee cherries when these are bright red, glossy and firm.
- ☆ Select the ripe cherries and pick them by hand. Ripe, unripe and over-ripe cherries are harvested by strip harvest. All types of beans can be harvested by machine also.
- ☆ To harvest the ripe beans, the coffee should be harvested by hand. Only 12-20 kg of coffee of good export quality can be collected from 100 kg of harvested cherries.
- ☆ After 3-4 years, the coffee tree bears fruit in lines or clusters along the branches of trees. The fruit turns red when it become ready for harvesting. Coffee beans are seeds of these ripened cherries. Arabica ripens in 6-8 months and Robusta takes 9-11 months.
- ☆ There are two beans which are covered by a thin membrane, this membrane in the coffee trade is known as "Silver Skin". Harvesting is done only once a year in the North of the equator during September- March and twice in the South of the equator during April-May and continues up to August.
- ☆ The average coffee tree produces 0.45- 0.7 kg of roasted coffee beans per year.
- ☆ Then, it can be prepared either by a "dry" method, which produces natural coffee or by a "wet" method, which produces washed coffee.

Insects

- ☆ **Coffee berry borer:** The insect bore into berries and causes economic loss to the growers.

A collective approach should be taken by all growers. Control should start just after the harvest of the crops to reduce the colonies. Use alcohol trap @ 22 trap/ ha.

- **Coffee stem borer:** It makes a hole in the stem and a whitish coloured saw is seen near the base of the stem. It reduces normal growth and plants may die.

  Manual killing of insects by inserting wire in the hole. Paint the stem with insecticide, it kills eggs and larvae. Inject the insecticides or Kerosene oil in the hole. After that, cover it with mud or cotton.

- **Coffee leaf miner:** It sucks the sap from the leaf. Nobel weed may be kept on boundaries or the roadside for the survival of beneficial insects.

  Use pheromone traps or apply botanical pesticides.

- **Coffee mealybug:** It excretes a honeydew like substance, which leads to the formation of sooty mould.

  It causes defoliation, fruit drops and dying of branches. Apply malathion or quinalphos @ 1.5 per cent around the stem and under the shade of the bush.

- **Coffee root mealybug:** It feeds on the root of the plant and waxy material is the most important sign of root mealybug infestation.

  Use clean media. Don't allow water to flow from the infested field. Submerge the pot of coffee plant in hot water at 120°F temperature. Apply neem cake.

- **Coffee scale insect:** It sucks the sap from leaves and shoots. It excretes honeydew, which results in the formation of sooty mould fungus and leaves curl upward.

  Remove the host plant. Spray *Verticillium lecanii* fungus @ 6 x 10^6 spores/ ml. Release Australian lady bird beetle (*Cryptolaemus montrouzieri*) @ 750 ml/ ha.

Diseases

- **Leaf rust:** A yellowish coloured spot appears on the lower side of leaves after the first rains of the season particularly in Arabica coffee. Spot size increases with the advancement of disease and causes defoliation.

  Spray Bordeaux mixture @ 0.5 per cent during pre and post-monsoon.

- **Black rot:** It occurs in high rainfall and high humidity areas during monsoon. Blackening and rotting of leaves, twigs and berries of affected plants.

  Spray Bordeaux @ 1.0 per cent or carbendazim @ 0.03 percent.

- **Coffee berry disease:** Infected berries have dark sunken spots and these spread and cover whole berries.

  Spray Bordeaux @ 1.0 per cent.

- **Collar rot/ damping off:** Water-soaked spots occur during the nursery stage at collar regions. The seedling may die during the severity of the disease.

  Drenching of rooting area with copper oxychloride @ 0.25 per cent.

- **Die back or anthracnose:** Circular greyish spots appear on the berries, and beans become brown. In a later stage, the plant may die.

  Spray mancozeb @ 0.25 per cent.

Chapter 28

Rubber (*Hevea brasiliensis* Muell-Arg.)

Family: Euphorbiaceae; 2n = 36

Introduction

The latex of the plant is used for commercial or industrial purposes. Plants start yielding latex at the age of about 3 years. It is used for the preparation of different items like tyres, tubes, tools handle, interior decorations, window components and other products. Only the tyre industry consumes about 85 per cent of the total rubber production and its consumption is increasing at a fast rate. In India, this crop was introduced in Kerala in 1902.

Rubber Research Institute of India, Kottayam, Kerala.

Origin

Rubber crops originated in the Amazon River basin of Southern America. This crop was introduced to South Asia through Kew Garden in the U.K in late 1970's. Now, it is grown in tropical regions of Asia, Africa and America.

Soil and Climate

It requires deep and lateritic fertile soil with an acidic pH of 4.5 to 6.0 and is highly deficient in available phosphorous. A tropical climate with an annual rainfall of 2000 to 4500 mm is suited for its cultivation. The ideal temperature for cultivation of rubber should range from 25 to 34°C, with 80 per cent relative humidity. Regions prone to heavy winds should be avoided for growing rubber crops.

Varieties

Tjir 1, PB 86, BD 5, BD 10, PR 17, GT 1, RRII 105, RRIM 600, PB 28/59, PB 217, PB 235, RRIM 703, RRII 5, PCK-1, 2 and PB 260.

Propagation

Seed Propagation: Seeds are sown under shade in the sandy loam soils. The seed starts germinating in 6-10 days and is completed in about 3 weeks. Seeds are germinated in shaded beds of friable soils, sand or coir dust. Germinated seeds are transplanted into shaded nurseries at spacing depending on the type of plant required.

Fig. 28.1: Tapping in Rubber

Clonal Seeds

The sexual progeny of budded clones is known as clonal seed. Seed garden progeny is probably a better name. Clonal seedlings have more variability than budded rubber, their average yield is also less. Legitimate seeds may be produced by hand pollination between selected clones.

Budding

Take bud from a selected mother plant. Seedling is budded when it attains pencil thickness above the collar region. Budded seedlings are headed back/ stumped 10-15 cm above the bud union about 4 weeks after budding and transplant in the field.

Green budding with buds stripped from green shoots is also practised.

Table 28.1: Green Budding and Brown Budding in Rubber

Sr.No.	*Particulars*	*Brown budding*	*Green budding*
1	Age of stock.	> 10 months old.	2 to 8 months old vigorous seedlings.
2	Maturity or age of scion or bud wood.	One year.	6 to 8 weeks.
3	Colour of bud wood.	Brown buds are in the axils of fallen leaves.	Green buds are in axils when leaves are still attached and functional.

Nursery

Bed size: 60-120 cm width of convenient length.

Spacing: For seedling stumps: 23 cm x 23 cm, 30 cm x 30 cm and 34 cm x 20 cm.

Budded stumps: 30 cm x 30 cm (budding the clone on seedling *in-situ*).

Stump budding: 60 cm x 60 cm (rootstock is cut as stump and budding is done).

Bud wood nursery: 60 cm x 90 cm or 60 cm x 120 cm.

Season

June – July is an optimum time for propagation and planting.

Seedling Nursery

Manuring

- ✰ Apply farmyard manure @ 2.5 t/ha and rock phosphate @ 350 kg/ha as basal dose 1.5 to 2 months after planting.
- ✰ Apply urea @550 kg/ha after 3 to 3.5 months.
- ✰ For immature rubber trees at the pre-tapping stage.
- ✰ Apply 12 kg compost or farmyard manure and 120 g rock phosphate in each pit before planting. Apply 10:10:4:1.5 ratio of NPK and Mg mixture as per the schedule given below:

Table 28.2: Application of Fertilizers

Months after planting	*Period of application*	*Quantity per plant*	
		10:10:4:1.5 (NPK Mg)	*12:12:6 (NPK)*
3	September/October	225 g	190 kg
9	April/May	450 g	380 kg
15	September/October	450 g	380 kg
21	April/May	550 g	480 kg
27	September/October	550 g	480 kg
33	April/May	450 g	380 kg
39	September/October	450 g	380 kg

☆ Apply 400 kg of mixture/ha in 2 doses, 1st during April/May and 2nd during September/October from the 5th year till the tree is ready for tapping.

Planting

A pit size of 1 m^3 is dug and filled up with top soil and compost in 1:1 ratio.

Table 28.3: Spacing and Plant Population

Planting material	*Spacing (m)*	*Population/ha*
Budded plants		
Hilly areas	6.7 X 3.4	445
Plains	4.9 X 4.9	420
Seedlings		
Hilly areas	6.1 X 3.0	539
Plains	4.6 X 4.6	479

In situ sowing: Germinated seeds are sown *in situ* in the pits. Healthy ones are retained and the others are removed.

Matured Rubber Trees under Tapping

For mature rubber trees under tapping, apply NPK 10:10:10 grade mixtures at the rate of 900 g/tree (300 kg/ha) every year in two split doses. Add 10 kg commercial magnesium sulphate for every 100 kg of the above mixture if there is a magnesium deficiency.

Cover Cropping

Grow crops to cover the ground under rubber crops and to maintain soil productivity. The most common crops that can be grown under rubber are;

1. ***Pueraria phaseoloides*:** Popular cover crops in India.
   a. Vigorous twiner and creeper that can be propagated by seeds and cuttings.
   b. It can stand against strong sunlight and smother weeds.

   Seed rate: 3 to 4.50 kg per ha.

   Other cover crops are:
2. ***Calopogonium mucunoides*:** A twiner and creeper die off during dry months.
3. ***Centrosema pubescens*:** Climber can be grown in shade. However, it cannot be grown under wet conditions.

Mulching

☆ Young plantations should be weeded completely to avoid weed competition.

- ☆ Mulching with dry leaves, water weed (*Salvinia* spp), grass cuttings, cover crop, lopping, *etc.* It prevents soil degradation around the plant.

Aquatic weed

- ☆ *Salvinia spp.* (African payal) is spreading as a menace in the waterways and it can be applied @ 5 kg/m^2 as sun-dried material.

Season of mulching

- ☆ After monsoon *i.e.*, November.

Weed Control in Rubber

- ☆ Regular weeding is required till the establishment of a cover crop, it takes about 2 years. There will be little or no growth of weeds after the establishment of a cover crop.

Table 28.4: Duration and Intensity of Tapping

Duration of tapping		*Intensity of tapping*
S/2, d/2	Half spiral, alternate day for 6 months and rested for 3 months.	100 per cent intensity.
S/2, d/26 m/9	Half spiral, alternate day for 6 months and rested for 3 months.	67 per cent intensity.
S/2, d/3	Half spiral, third day.	67 per cent intensity.
S/2, d/31 m/2	Half spiral, daily for one month and rested for next month.	100 per cent intensity.
S/1, d/4	Full spiral, fourth day.	100 per cent intensity.
V/2, d/2 12 m/16	Half circumference and cut alternate days for 12 months and rested for the next 4 months.	75 per cent intensity.

Ethrel Treatment

Apply ethrel @ 0.5 per cent a.i. 5 cm below the tapping cut after scrapping the outer bark. The first application is made just after a few pre-monsoon showers and thereafter in the month of September and November. Ethrel should not be applied more than thrice per year.

Tapping Panel Dryness (Brown blast)

It is characterized by partial or complete drying of the tapping area or starts declining (10-25 %) due to prolonged dripping.

Control: Give rest or reduce tapping intensity for 3 to 12 months.

Yield

- ☆ Rubber trees reach at its peak yield after 14 years of planting.
- ☆ In South India, the yield of seedling plants remains about 375 kg/ha, whereas the yield of budded plants is about 800 - 1000 kg/ha.

☆ Latex contains about 32 per cent dry matter.

Insects

☆ **Scale insect:** Small insects feed on leaves and stems. It sucks the cell sap.

Spray malathion 50 EC @ 2 ml/litre water just after the start of the infestation.

☆ **Mealybug:** The insect feeds on leaves and fruit in colonies. It gives a look like a cottony appearance. Both nymph and adult suck the sap.

Spray fish oil rosin soap 25 g/litre. Release Australian ladybird beetle (*Cryptolaemus montrouzieri)* @ 10 g/tree.

☆ **Termite (White ant):** Damage is more severe in newly planted orchards and nurseries. Formation of soil tunnels on plant and serrated cut seen on the cut portion of stem or root. It feeds on the root and lower portion of the stem by making tunnels of soil.

Apply or drench with chlorpyriphos 20 EC @ 2 ml/litre.

☆ **Cockchafer grub:** It feeds on the roots of young plants and seedlings. Affected plants droop and fall over. Control is same as in termite.

☆ **Mites:** It has four pair of legs and sucks the sap from the leaves and resulting in the shedding of leaves.

Spray sulphur 50 WP @ 2 g/litre or spray Dicofol 18.5 EC @ 2.5 ml/litre.

Diseases

☆ **Abnormal leaf fall:** A large number of infected leaves fall either green or turn coppery red. Lesions may develop on the petiole or leaf margin.

Remove infected and dried twigs. Spray Bordeaux mixture 1.0 per cent @ 4000-5000 litre/ha with a higher volume sprayer prior to the onset of monsoon. Spray 0.5 % Bordeaux mixture + 0.5 per cent zinc sulphate. Adopt an integrated approach like growing resistant variety, following proper cultural practices, using bioagents and apply recommended chemicals.

☆ **Secondary leaf fall:** Spray for abnormal leaf fall will also control secondary leaf fall.

☆ **Powdery mildew:** Leaves give ash coating appearance. Leaves roll inward and fall, but the petioles remain attached to the plant.

Apply sulphur powder @ 11-14 kg/ha after commencement of bud break before 10 am. Repeat it at 7-15 days intervals.

Sulphur powder can be mixed with talc powder in 7:3 ratio to control powdery mildew in nursery or young plants @ 1kg/4000 litre water.

☆ **Bird's eye spot:** A circular spot appears on the leaf with a dark centre and brown boundaries. It is most common in nursery and young plants.

Spray Bordeaux mixture 1 per cent or Mancozeb or copper oxychloride @ 0.2 per cent. Timely application of nutrients or organic manures is also mandatory for tree vigour and proper growth of plants.

- **Leaf spot:** Dark brown scattered spots appear on the leaf. Later, lesions decay and the margin looks like a black ring.

  Apply following fungicides at 15 days intervals. Spray Bordeaux mixture @ 1.0 per cent or Mancozeb @ 0.2 per cent or Carbendazim @ 0.1 per cent to control leaf spot.

- **Pink disease:** Appearance of sunken and swollen spots on stem and branches (pink pustules). It may cause splitting of bark after the onset of monsoon.

  Regular supervision is required from July to September to find the infection and apply Bordeaux paste at least 30 cm above and below the infected area. Prune off or remove the bark of the infected area and burn the removed bark. Apply Bordeaux paste on the bark removed portion of the plant.

- **Patch canker or bark cankers:** The area under the tapping cut turns brown and rot, which reduces latex production as well as latex flow.

  Remove the bark of the infected area along with coagulated rubber. Wash the wound with Dithane M 45 @ 0.75 per cent. After drying the fungicide, apply the wound dressing compound.

- **Dry rot, Stump rot, Collar rot or Charcoal rot:** Affected bark becomes soft and powdery, and a double black line appears on the wood.

  Remove the bark of the affected area and wash it with carbendazim 0.1 per cent solution. If the affected area shows black lines, wash it with a fungicide solution. After drying the washed portion, apply the wound dressing compound. Avoid accumulation of rubber on the affected portion.

- **Brown root disease:** The formation of dark brown crust by fungus (*Phellinus noxius*). Affected roots start rotting.

  Expose the root area and remove the dried roots. Wash the partially dried and infected/unhealthy roots with carbendazim 0.1 per cent solution. Apply wound dressing compound after drying the fungicide solution from the infected area. Cover the exposed area with soil and drench with fungicide solution.

Appendix I

List of NRCs/Institutes/Directorates/Universities Exclusively Working on Horticulture

National Research Centers (NRCs)

- ☆ ICAR-National Research Centre for Banana, Trichi, Tamil Nadu
- ☆ ICAR-National Research Centre for Grapes, Pune, Maharashtra
- ☆ ICAR-National Research Centre for Litchi, Muzaffarpur, Bihar
- ☆ ICAR-National Research Centre for Pomegranate, Solapur, Maharashtra
- ☆ ICAR-National Research Centre of Orchids, Pakyong, Sikkim
- ☆ ICAR-National Research Centre on Plant Biotechnology, New Delhi
- ☆ ICAR-National Research Centre on Seed Spices, Ajmer, Rajasthan

ICAR Institutes

- ☆ ICAR-Indian Institute Horticultural Research, Hessaraghatta, Bengaluru, Karnataka
- ☆ ICAR-Central Institute for Arid Horticulture, Bikaner, Rajasthan
- ☆ ICAR-Central Institute of Sub Tropical Horticulture, Lucknow, UP
- ☆ ICAR-Central Institute of Temperate Horticulture, Srinagar, J & K
- ☆ ICAR-Central Institute on Post Harvest Engineering and Technology, Ludhiana, Punjab.

- ICAR-Central Plantation Crops Research Institute, Kasargod, Kerala
- ICAR-Central Potato Research Institute, Shimla, Himachal Pradesh
- ICAR-Central Tobacco Research Institute, Rajahmundry, Andhra Pradesh
- ICAR-Central Tuber Crops Research Institute, Trivandrum, Kerala
- ICAR-Indian Institute of Spices Research, Kozhikode, Kerala
- ICAR-Indian Institute of Vegetable Research, Varanasi, UP
- ICAR- Indian Institute of Oil Palm Research, Pedavegi, West Godawari, Andhra Pradesh
- ICAR-Central Citrus Research Institute, Nagpur, Maharashtra

Directorates

- ICAR-Directorate of Cashew Research, Puttur, Karnataka
- ICAR-Directorate of Medicinal and Aromatic Plants Research, Anand, Gujarat
- ICAR-Directorate of Floricultural Research, Pune, Maharashtra

Universities

- Maharana Pratap Horticulture University, Karnal, Haryana
- Dr. Yaswant Singh Parmar University of Horticulture and Forestry, Solan, Himachal Pradesh
- University of Horticulture Science, Bagalkot, Karnataka
- University of Agriculture and Horticulture Sciences, Shimoga, Karnataka
- Sri Konda Laxman Telangana State Horticultural University, Hyderabad, Telangana
- VCSG Uttarakhand University of Horticulture and Forestry, Bharsar, Uttarakhand

Source of Database

www.nhb.gov.in

https://www.fao.org/faostat/en/

https://apeda.gov.in/

https://agricoop.nic.in/en/statistics/publication-reports

www.hortharyana.gov.in

https://www.indiastat.com/Home/Login

https://icar.org.in

Appendix II

Area, Production and Productivity of Various Horticulture Crops in India (2021-2022)

Crop	*Area (In ' 000 hectare*	*Production (000' MT)*	*Productivity (In MT/hectare)*
Fruits			
Almond	10	12	1.17
Aonla/Gooseberry	105	1278	12.21
Apple	315	2589	8.22
Banana	963	34528	35.87
Ber	52	547	10.61
Citrus			
(i) Lime/Lemon	313	3776	12.08
(ii) Mandarin	454	6303	13.88
(iii) Sweet Orange (Mosambi)	228	3895	17.06
(iv) Other Citrus	103	782	7.59
Citrus Total (i to iv)	**1098**	**14757**	**13.44**
Custard apple	48	418	8.69
Grapes	163	3401	20.81
Guava	354	5527	15.62
Jackfruit	195	3301	16.94

Crop	Area (In '000 hectare	Production (000' MT)	Productivity (In MT/hectare)
Kiwi	5	16	3.11
Litchi	99	737	7.42
Mango	2350	20772	8.84
Muskmelon	68	1498	21.87
Papaya	150	5341	35.71
Passion Fruit	11	56	4.93
Peach	19	121	6.33
Pear	44	309	7.05
Picanut	1	0	0.54
Pineapple	108	1851	17.06
Plum	24	86	3.62
Pomegranate	270	3086	11.41
Sapota	78	941	12.07
Strawberry	3	11	4.43
Walnut	104	289	2.79
Watermelon	120	3308	27.58
Other Fruits	306	2726	8.90
Total Fruits	**7064**	**107507**	**15.22**
Total Vegetables	**11374**	**209143**	**18.39**
Aromatics and Medicinal	668	664	0.99
Flowers Cut	-	615	-
Flowers Loose	282	2199	7.79
Total Flowers	**282**	**2813**	**9.96**
Plantation Crops			
Arecanut	849	1666	1.96
Cashewnut	1184	752	0.64
Cocoa	106	28	0.27
Coconut	2154	13317	6.18
Total Plantation	**4293**	**15764**	**3.67**
Total Spices	**4360**	**11155**	**2.56**
Total	**28042**	**347179**	**12.38**

Source : Ministry of Agriculture and Farmers Welfare, Govt. of India. (ON3224)

Index

9 789359 197302

Printed at Repro India Ltd.